Modeling with Mathematics

Authentic Problem Solving in Middle School

NANCY BUTLER WOLF

Foreword by **Max Ray-Riek** of The Math Forum

Heinemann
Portsmouth, NH

Heinemann
361 Hanover Street
Portsmouth, NH 03801–3912
www.heinemann.com

Offices and agents throughout the world

Cover image: © Jeremy Woodhouse/Space Images/Corbis

Interior images: © Getty/iStock/boboling; © Getty/iStock/akiyoko; © Getty/iStock/boboling; © Getty/iStock/Bee_W; © Digital Vision Vectors/mstay; © Getty/iStock/cm1; © Getty/iStock/Eskemer; © Getty/iStock/Kyoshino; © Getty/iStock/Eskemer; © Getty/iStock/belchonock; © Getty/iStock/captainweeraphan

Library of Congress Cataloging-in-Publication Data
Wolf, Nancy Butler.
    Modeling with mathematics : authentic problem solving in middle school / Nancy Butler Wolf.
        pages cm
    ISBN 978-0-325-06259-4
    1. Mathematics—Study and teaching (Middle school).   I. Title.
QA135.6W63 2015
510.71'2—dc3   2015028825

Acquisitions editor: Katherine Bryant
Production editors: Sonja S. Chapman and Patty Adams
Cover design: Monica Ann Crigler
Interior design: Catherine Araklian
Typesetter: Gina Poirier Design
Manufacturing: Veronica Bennett

Printed in the United States of America on acid-free paper
19 18 17 16 15 VP 1 2 3 4 5

Dedication

To the most important educators in my life:

The past

My parents, who taught me through word and deed to never stop learning.

The present

My partner, Mimi, who is a superior educator, and teaches me every day about life, love, and laughter.

The future

My deepest pride and joy, my daughters Erin and Alexis, who have followed in the family trade. I could not be prouder that they have chosen to teach. I am confident they will touch young lives in a profound way.

Contents

Acknowledgments vii

Foreword ix

1 **Modeling with Mathematics** 1
What Is Mathematical Modeling? 3
The Process of Modeling 7
Why Modeling? 10
Mathematical Autonomy 12

2 **Rich Mathematical Modeling Tasks** 16
Characteristics of Rich Modeling Tasks 16
Word Problems Versus Modeling Tasks 22
Finding Rich Mathematical Modeling Tasks 23

3 **Identifying and Investigating Problems** 33
Getting Started: Student Autonomy 34
Getting Started: The Teacher's Role 37
Identifying More Problems: Old MacDonald's Farm 40
Getting Started: 1,000 Paper Cranes 41
Conclusion 45

4 **Formulating the Model, Collecting Data, and Finding a Solution** 46
Formulating a Model 47
Using Multiple Representations 49
Formulating the Model and Collecting Data:
        The 1,000 Paper Cranes Problem 63
Conclusion 70

⑤ Interpreting the Solution and Comparing with Reality **72**

The Importance of Interpretation 73

Interpreting the Solution. 75

Examples of Students Interpreting Solutions 84

Conclusion 88

⑥ Communicating and Implementing the Solution **89**

Communicating the Solution 90

Using Multiple Representations. 101

Implementing the Solution. 105

Working with the 1,000 Paper Cranes Problem 106

Conclusion 108

⑦ Assessing Mathematical Modeling. **109**

Goals of Assessment. 110

Informal Assessment 110

Formal Assessments. 116

1,000 Paper Cranes Assessment. 125

Conclusion 128

⑧ Doing Mathematics. **129**

Works Cited **137**

Online Resources

The Road Trip: A Mathematical Modeling Task

The Vegetable Garden: A Fixed Perimeter Investigation

Setting Up a Budget: A Modeling Application Problem

Kicker Ramps

1,000 Paper Cranes

The Block Tower Problem: An Introduction to Mathematical Modeling

Old MacDonald's Farm

The Tortoise and the Hare: A Rate, Time, Distance Investigation

Barbie Bungee

Mixing Paint: Finding the Goof

The BMX Bike Race

School District Pay Raises

The Giant's Footprint

The Thanksgiving Dinner Party: An Algebraic Thinking Investigation

The Pasta Box Task: A Geometric Modeling Investigation

To access the online resources:

Go to http://heinemann.com/products/E06259.aspx and click on
Companion Resources for downloadable modeling task sheets.

Acknowledgments

My deepest gratitude goes out to all those who have led me, encouraged me, and walked with me on this journey:

My family, who have encouraged my work and been patient with my time. My mom has shown constant enthusiasm and confidence in my ability to complete this task. My daughters have been encouraging and patient with my obsession about this work. My partner, Mimi, has listened to every word I have written, offered suggestions and praise, and brought unwavering support, humor, and cappuccinos.

My Heinemann editor, Katherine Bryant, who has walked me through every step of this new endeavor with patience, wisdom, and positive energy. She is a master at her craft, and has helped transform a whirlwind of ideas, lessons, and words—and a passion to share them—into a book.

My CGU colleagues and professors, especially Dr. Gail Thompson and Dr. David Drew. Dr. Thompson encouraged me early in my PhD program—when I was lacking confidence and feeling like a bit of an imposter—to write my own book. David has truly been my greatest mentor. His wit, wisdom, and confidence in me have sustained me throughout this journey. He is the carver who helped me step forth.

My Charter Oak colleagues, who have supported my work and been generous with their classrooms and their students.

My ITUM compadres Lisa Maggiore, Bob Colera, Nitza Peraza, and Dr. Judith Jacobs. My work (and play) with all of you lit a passion in me for math education. I have learned from you not only about mathematics and teaching, but also about collaboration, friendship, and laughter.

My teachers. So much that is good and positive and enriching in my life has been because of teachers. From my earliest school days through my last classes in graduate school, I have been blessed with kind, loving, intelligent, and encouraging teachers. You have shown me the wonder and beauty of learning and the power of a teacher in a person's life. I am proud to be counted as one of you.

My students. I have learned more from you than I could learn in a lifetime of research and study. I do what I do to make this world a better place for you.

Foreword
by Max Ray-Riek

Thanks to the Standards of Mathematical Practice in the Common Core State Standards, mathematical modeling is a hot topic among math teachers. One reason it has inspired so much discussion is because it is probably the mathematical practice that is *least* a part of our usual routines. All math classes involve making sense of problems and persevering in solving them; most involve constructing viable arguments; using tools strategically and reasoning abstractly and quantitatively come with the territory. But what does modeling with mathematics even mean? Does it require radical shifts in practice, or worse, in lesson planning?

Mathematical modeling has not been a part of the K–12 classroom in part because it's not even part of most teachers' math preparation. As a high school teacher, I was required to major in mathematics. However, I didn't have to take any classes in which we did modeling—and I avoided them because I was scared of modeling! I was worried that courses in applied mathematics would require lots of scientific content knowledge that I didn't have, involve lots of computation with confusing technology, and nearly infinite math knowledge—after all, the world is so complex, wouldn't you have to know pretty much everything to build a reasonable model?

So with teachers who haven't had experience creating mathematical models in their own education, and in classrooms where little modeling has taken place, the idea that one eighth of the Standards of Mathematical Practice is devoted to modeling seems frightening. For example, you might be reading this foreword and thinking, "I don't know what modeling even is, and it already sounds intimidating."

Luckily, Nancy Butler Wolf has written a book that reads like an experienced teacher opening the door to her classroom and handing over all of her favorite lessons and rubrics, complete with her own stories of what worked—and didn't work—as she learned to teach using mathematical modeling.

As you read the book, you will likely come to see, as I did, that it is about using good, rich problems that encourage students to make sense of problem situations, research or generate data, organize their thinking using multiple representations, include grade-level math content, and then share their thinking in a way that ensures that they are making sense, reflecting, connecting, and justifying. That doesn't sound so bad; it sounds like the mathematics we want students to be doing, and the mathematics we do in our daily lives as we think about real situations ranging from "Does it make financial sense to buy a more expensive, hybrid car?" to "Is it going to be possible to fit the washing machine we want down our basement stairs?"

Each chapter has at least one example of a task that invites students to use middle school mathematics (linear relationships, solving equations, proportional reasoning, volume, surface area, etc.) to answer engaging questions about real or imaginary scenarios. The book also tells the story of a single rich task, designing a display of 1,000 paper cranes, from its inception (as part of a cross-curricular project) to the final presentations to the principal, including how the students were scored with a rubric, not to mention photographs of the completed scale models of the crane projects! Between the stories, the examples, the extensive citing of research, and the way Nancy describes her own transformation as a teacher, I'm confident that you'll be excited and ready to try modeling tasks with your students (or improve your current teaching with rich tasks) as soon as you finish this book. What are you waiting for?

Modeling with Mathematics

1

No problem can withstand
the assault of sustained thinking.

—VOLTAIRE

Several years ago, I had a lightbulb moment about teaching and learning
mathematics: mathematics devoid of meaning is empty. I had taught a unit
on factors, multiples, and divisibility, and all indications pointed to the fact that
the students understood this concept. Every check for understanding showed
me that students could find all the factors of given numbers in a systematic way,
and could use divisibility rules to determine factors quickly and easily. Then I
gave them a traditional, formal assessment. Nearly every student did well finding
factors and multiples, assessing divisibility, finding the Greatest Common Factor
and the Least Common Multiple. Then I put the last question on the assessment:

> There are 96 students in the school band. The director wants to
> arrange the students into rows, each containing the same number
> of students, with no students left over. Can she arrange them into:
>
> > Rows of 2?
> > Rows of 3?
> > Rows of 4?
>
> Are there any other ways she could arrange the students without
> any students being left over? Describe them below.

I had imagined this would be a fairly simple application problem for the students who clearly understood factors and divisibility. Boy, was I wrong! I was astounded to find that most students (including those who answered every other question correctly) could not correctly answer this question. I learned some lessons from this experience:

1. Although our students may demonstrate competency with mathematical concepts, if they cannot apply their knowledge, they do not truly have an understanding of the algorithms they are performing or the rules they are utilizing.

2. Most traditional math books emphasize repeated calculations with, perhaps, a few word problems thrown in at the end of the section or the unit.

3. As much as we as teachers might explain the meaning of various concepts and provide formal definitions, students must be given frequent and varied opportunities to apply their knowledge.

4. The failure was mine.

I had been patting myself on the back when teaching this unit, certain that the students understood the mathematics. I had certainly discussed the meaning of *factor, multiple, divisible,* and other terms used in the work we were doing. I had even engaged students in sense-making discussions when working on problems (Student: "The factors of 24 are 1, 2, 3, 4, 6, 8, 12, and 24." Me: "Good. What does it mean that those are the factors? How did you find the factors? How do you know there are not other factors?"). However, I had not provided students with enough opportunity to apply the concepts while talking and writing about them. This experience began the transformation in my teaching and my education about teaching and learning.

Since that day, with lesser and greater success, I have sought to provide my students with real, meaning-making applications for the math concepts I am presenting. It has been my joy to see students for whom mathematics had always been a source of confusion, frustration, and failure learn how math is connected to their everyday lives and that it can be fun and engaging at the same time. Rather than merely performing calculations, it is essential to provide students with opportunities to model real-life problems and solve those problems in a meaningful way using a variety of strategies.

A great divide often exists between students' conceptual understanding and their procedural skill (Hiebert et al. 1996; Lesh 1981). The traditional course of action is to learn a concept, perform calculations on multiple items related to that concept, learn to solve a few problems using that concept, and then, perhaps, apply the knowledge. The belief that application is complex and complicated often leads teachers to stop short of this last, most important step. Students do not best learn in this manner. Problem solving, modeling, and application must be embedded throughout the process of student learning (Lesh 1981; Weber et al. 2010).

What Is Mathematical Modeling?

Mathematical modeling is a powerful tool and skill for mathematical learners. The National Council of Teachers of Mathematics (NCTM) has identified mathematical modeling as one of its major focal points in algebra standards across the grade levels. From pre-K through grade 12, students are expected to "Use mathematical models to represent and understand quantitative relationships" (www.nctm.org/Standards-and-Positions/Principles-and-Standards /Algebra/). The concept of mathematical modeling is also a major emphasis in the Common Core Standards for Mathematics (CCSSM). The CCSSM identified eight Standards for Mathematical Practice that define the processes that math educators are working to develop in all students grades kindergarten through 12. One of these practices is modeling with mathematics. This practice emphasizes a student's ability to apply mathematical tools to solve real-life problems and to analyze his or her solution to determine whether it makes sense in context (National Governors Association Center for Best Practices 2010).

➤ A Shift in Thinking

The new focus on mathematical modeling will require a shift in thinking and practice for the nation's math teachers. For many, this will be a major shift. During the past couple of decades (the teaching life of the majority of math teachers in the United States), the focus in classrooms has been success on standardized assessments. By 2001, with the No Child Left Behind Act, the pressure was on for teachers, schools, and districts to demonstrate students' understanding of standards through a series of high-stakes tests. These same teachers, schools, and districts were held accountable for their students' test results and faced consequences ranging from interventions and corrective actions to school restructuring for poor performance (Federal Education Budget Project 2012).

Because of the pressure to perform on high-stakes, multiple-choice tests, many schools and teachers used their class time to "cover the standards" and prepare students for testing. This preparation left little time for true problem solving, investigation, or modeling tasks. I have presented problem solving and mathematical modeling problems at professional development sessions and conferences for several years. The two most common concerns expressed by teachers are some variation of "How can I do this project in my classroom and still get through Chapter 12?" and "How will this project help my students do well on the state tests?" Now that the tide is turning away from multiple-choice exams, and the focus of mathematics education is turning toward problem solving, real-life application, and critical thinking, teachers are anxious to change the way they teach.

Teachers understand that students' motivation and engagement in math are increased when the students are engaged in meaningful problem solving (Deci and Ryan 1982; English, Fox, and Watters 2005), and they are excited about the possibilities for their own classrooms. In a recent survey of more than 300 kindergarten through twelfth-grade math teachers, nearly 95 percent responded that they were willing or eager to change their teaching practice in response to the Common Core Standards; more than 92 percent believed that mathematical modeling with be beneficial to their students; and nearly 86 percent reported that they are excited to try new lessons using mathematical modeling (Wolf 2013b). It is important, then, that teachers understand what mathematical modeling is (and what it is not).

➤ Word Problems, Problem Solving, and Modeling

In my work and discussion with teachers, I have discovered a common misconception about mathematical modeling. Many teachers view mathematical modeling as a process of showing the students how to approach or solve a problem. The key feature of mathematical modeling, as defined by the NCTM and the CCSSM, is modeling *with* mathematics. So, although the teacher is obviously the facilitator of the process, the modeling is done primarily by the students.

Mathematical modeling involves the work and stages included in identifying a problem, choosing the tools needed to solve the problem, implementing these tools to find a solution, and testing that solution to determine whether it makes sense in the context of the problem situation (Swetz 1991). Modeling can be differentiated from word problems in several ways. Word problems in most mathematics classrooms, while more demanding than pure computation problems, are typically presented in the context of a specific math content area, and are solved with a particular method or algorithm. Modeling, by contrast, does not usually call for the use of one method or algorithm in order to solve the problem. Also, standard word problems in school mathematics curricula do not model realistic problem situations and problem solving for students, whereas modeling presents students with realistic problem-solving experiences requiring strategizing, using prior knowledge, and testing and revising solutions in a real context (Greer 1997; Lesh et al. 2003; Verschaffel and De Corte 1997). As Richard Lesh (1981) says, "Among the word problems that are found in mathematics textbooks, very few might actually occur in a sane and reasonable life, and virtually none deal with real or even realistic data" (249).

Problem solving is typically viewed as a step above word problems, and mathematical modeling is a specific type of problem solving. Problem solving requires students to interpret what is needed to solve the problem and then determine how to find the answer. One distinct difference between typical problem solving and mathematical modeling is that modeling frequently involves interpretation or analysis of an essentially nonmathematical scenario. Students must study the scenario to determine what the important factors or variables are, interpret these mathematically, and develop a model. They then use the model they devise to analyze the problem situation mathematically, draw conclusions, and assess them for reasonableness of the solution (Swetz, Hartzler, and the National Council of Teachers of Mathematics 1991).

The following Problem Pyramid might be a good visual descriptor:

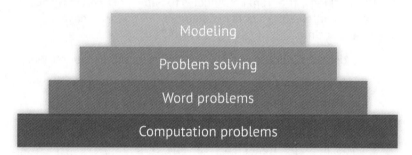

The pyramid illustrates the fact that the basic, and most frequently used, type of problems are computation problems. On Webb's (2002) Depth of Knowledge (DOK) Levels, these problems would represent Level 1 problems. That is, they usually require students to carry out a procedure or apply an algorithm. Word problems are typically used less frequently than pure computation problems, and might fall into a Level 2 DOK. At this level, students must demonstrate some conceptual understanding, and these problems require them to make some decisions on how to approach the problem. Problem-solving opportunities require more reasoning and strategic thinking than word problems. These problems require more critical thinking and the ability to explain reasoning and justify procedures. They would be classified as Level 3 DOK problems. Finally, students attempt mathematical modeling tasks less frequently than the other types of problems because they demand that students engage in extended thinking and complex reasoning. Students may be required to relate mathematical ideas to other situations and to real life. These experiences might be categorized as Level 4 DOK tasks.

From Word Problems to Modeling

There is a place in the classroom for each type of problem, but it is possible to take a typical word problem and adapt it in such a way that it increases the depth of knowledge required to solve it. For instance, a word problem at the end of a lesson on rate $\times$ time = distance might be the following:

> Mr. Wu makes a driving trip from one city to another at an average rate of 55 miles per hour. If the distance between the cities is 375 miles, how long would the trip take?

This problem is designed to reinforce the "rate times time equals distance" concept, but requires primarily computation with little reasoning. A modeling problem might be designed to parallel the rate-time-distance problem shown, but also rely on students' decision-making skills, prior knowledge, and problem-solving skills. The modeling problem might look like this:

> Mr. Wu is taking a trip from Los Angeles to San Francisco. He wishes to see some interesting sights in California, but does not have a great deal of time.
>
> Use the California map to plot 2 different routes for Mr. Wu.
>
> Assume the average rate on the various roads is as follows:
>
> 60 miles per hour on Interstate Highways
>
> 45 miles per hour on U.S. Federal Highways
>
> 30 miles per hour on State Roads
>
> For each route, determine the number of miles Mr. Wu will travel and the time it will take him. Show your calculations and explain your thinking.
>
> Then, write a letter to Mr. Wu explaining why one of your chosen routes is the best for him. Include information about the miles he will travel, the estimated time the trip will take, and the sights he can see along the way.
>
> Extension: Determine the cost of Mr. Wu's trip, including gas charges (at the current price of gas), price of entrance to any attractions or sights along the route, and lodging if required.

This open-ended problem allows students to use tools and prior knowledge including measurement, proportions, map reading, scale drawings, and geography to make decisions and justify those decisions in a real-life problem. Students may use multiple methods to come to a conclusion and multiple representations to demonstrate their understanding. They must explain their solutions and use writing in mathematics to explain their reasoning.

The Process of Modeling

The process of mathematical modeling is often broken down into a series of steps. There are several variations on the process, but some of the most common involve the following stages:

1. Investigation and Problem Identification

2. Mathematical Formulation of the Model and Data Collection

3. Obtaining a Mathematical Solution to the Model

4. Interpreting the Solution and Comparing to Reality

5. Communicating and Implementing the Solution

(Swetz 1991; Swetz, Hartler, and National Council
of Teachers of Mathematics 1991; Warwick 2007)

Let's look a brief walk-through of the Road Trip problem in light of the stages of modeling. (We will look at other problems in more detail later!)

➤ Investigation and Problem Identification

In this stage, students consider all the information in the problem and decide what is most important, what is less important, and what is unnecessary. They also begin to make decisions about materials they may need and some methods they may use to solve the problem. They access prior knowledge necessary for their work and decide what they still need to investigate.

For this stage of the Road Trip problem, students work in small groups and might begin by looking at a map to find Los Angeles and San Francisco (a little easier for my California students!). They consider the major routes between the two cities, and begin to familiarize themselves with the various road markers on the map. They may look at the map scale and discuss major sights with which they might be familiar along the route. Students begin to compile and note some sightseeing preferences, and plan to investigate which sights are truly along the route of choice.

At this stage, students begin to identify the math skills and tools they will need to solve the problem. Students working on the Road Trip task will have some familiarity with proportions, and some experience solving basic proportions problems. They should recognize that part of this task will involve proportions and proportional reasoning. They will also have some familiarity and experience with the rate × time = distance concept, and understand it will be useful here.

When students have had time to identify and do some initial investigation of the problem, they benefit by restating the problem in writing. It will also be helpful for them to make notes about what they tools will need, what skills they need to know, and what they already know about the problem. If students are keeping a math journal, this is a good place to record thoughts and plans.

➤ Formulating the Model and Collecting Data

At this stage of the modeling task, students begin to collect data and formulate the mathematical model. They decide whether their initial plans regarding the needed math skills and tools are appropriate for the data they are collecting. They begin calculations and make decisions about what calculations to use, what data is important and unimportant, and how results of calculations will contribute to the final solution.

At this stage in the Road Trip problem, students decide on the final route of choice and mark it on the map. They begin measurements of various distances on the map (between sightseeing stops, on various types of roads), and make decisions about recording data. At this point, they might decide to record their data in a table, in a list, or on a picture. They may also begin setting up proportions using the map scale to determine actual distances.

➤ Obtaining a Mathematical Solution to the Model

This stage of a modeling task is the actual calculation stage. If the students have identified the problem properly, collected all necessary data, and correctly identified math methods needed to solve the problem, this stage should be pretty smooth sailing.

For this stage of the Road Trip problem, students calculate and sum up distances for the route of choice, using the scale for the map they are using. They then calculate the time needed to travel each leg of the trip (from sight to sight), using the average rates for each type of road traveled. They compile the information to come up with final solutions, and check the problem again to be certain they have answered all parts of the question.

➤ Interpreting the Solution and Comparing to Reality

While most traditional word problems stop at Stage 3, the solution, a vital part of modeling is the fourth stage: interpreting the solution and comparing it to reality. Too often, problems have little basis in real life (and therefore, little need to compare with reality), but modeling problems are, by design, real-life problems. Therefore, students must check their answers to be certain they make sense in the context of the problem.

In the Road Trip problem, students discuss the feasibility of their solutions. The first thing they must consider is whether their answer makes sense in terms of the distance between Los Angeles and San Francisco. They might check the distance on a basic Internet search and determine if the solution they got is reasonable. In this way they can check for errors in calculations, and some students may find that their chosen route is considerably longer than the "direct route." They must then decide if they need to recalculate the route, or if they need to change the route in light of the fact that Mr. Wu "does not have a great deal of time." At this point, students also need to decide (based on distance, time, and time spent at the chosen sights) whether Mr. Wu will need to spend the night somewhere along the route. If this is so, students are encouraged to find an appropriate stopping-point in the trip. If the problem is extended to include the cost of the trip, the results must be checked against Mr. Wu's money constraints.

➤ Communicating and Implementing the Solution

The final stage may take many forms, depending on the task. Students may communicate their solutions in a variety of ways. The might make a poster that presents their results with some explanations and representations. They may make a digital presentation or video. They might make a pamphlet or instruction guide. They may provide a written explanation or description of the problem and the solution (this may include a letter to the editor or the president of the company, a job proposal, an explanation written for future students who might investigate this problem, etc.). The possibilities for communication are endless, and are only limited by the specifics of the task and the creativity of the students.

The Road Trip problem asks the students to write a letter to Mr. Wu describing their recommendations and their results in terms of routes, sightseeing opportunities, and time needed to make the trip (and cost, if appropriate). Implementation of a task like this one does not have to mean the students go on a road trip. The letter to Mr. Wu, perhaps accompanied by a pamphlet describing the journey, can complete this final phase.

The time spent on a modeling task depends on the complexity of the task. Do not be deterred by the time investment necessary for the Road Trip problem. This is typically a two-period project (three, if you include class presentations), but the task can be modified to fit the time constraints of your classroom. We will investigate many modeling tasks throughout the course of the book, and you will see that some are relatively quick and easy while others are more in-depth investigations that will take more time to solve more complex problems using more math tools. In any case, the modeling experience can be a rich and rewarding one for you and your students!

Why Modeling?

Mathematical modeling is not just a type of word problem—it is a mathematical practice. Modeling can be infused throughout the math curriculum, and can be used in conjunction with many content areas or standards. Modeling represents a shift from *learning* math to *doing* math. The inclusion of modeling in the math classroom can increase student engagement, increase students' depth of understanding, and provide opportunities for investigation, contribution, and success for all learners. Students who are involved in problem solving and inquiry-based activities such as modeling develop a positive disposition toward mathematics (Carpenter et al. 1989; Hiebert et al. 1996). Students who engage in modeling in the math classroom have increased mathematical autonomy and flexibility in the ways they use mathematics.

➤ Engagement

A good modeling task engages students. When a task can tap into a student's innate sense of wonder about the world around him or her, that student becomes engaged in the problem-solving process. When we can pique the interest and curiosity of students through problems that have a basis in reality, we encourage those students to question, investigate, and problem solve (Ormell 1991). Students too often view what happens in the math classroom as completely removed from and irrelevant to the real world. Modeling bridges this gap and allows students to understand that situations occurring around them in daily life involve and require mathematics.

➤ Deep Mathematical Understanding and Flexibility

I have presented mathematical modeling problems to groups of teachers on many occasions. Even while they concede that students will be interested and engaged in modeling tasks, many express a concern about the time invested and whether that time will pay off in terms of understanding and performance on assessments. Research demonstrates that when students are engaged in rich modeling tasks, they develop powerful conceptual tools that increase their depth of understanding of a variety of mathematical concepts and improve their mathematical abilities (Boaler 2001; Kaput and Schorr 2008; Lesh and Lehrer 2003). While students who learn mathematics in a traditional fashion perform well on traditional, standardized assessments, they tend to do poorly on tasks that require them to apply the math concepts to real problems. Students who learn mathematics through a modeling lens are better able to perform on both traditional and nonroutine assessments (Boaler 2001).

➤ Confidence

Mathematical modeling is a particularly effective practice for students who have a history of poor performance on traditional mathematics tasks. By incorporating modeling tasks into the classroom, we recognize all students as important contributors to the decision making and investigation of the problem at hand. Because a good modeling task will be based in real-life experience, all students have the ability to make contributions based on their prior knowledge. Such tasks also emphasize and require a broader range of mathematical abilities than algorithmic exercises, and therefore allow, "a broader range of students to emerge as being exceptionally capable" (Lesh and Lehrer 2003, 113). Students who may have a history of poor performance in math when their abilities and understanding are assessed solely on narrowly defined tasks and assessments can demonstrate significant ability and potential when given the opportunity to problem solve in a real-life modeling context (Lesh, Zawojewski, and Carmona 2003). It is important for teachers to emphasize to students that any (mathematically valid) solution for which they can make a strong argument is "correct." This flexibility in thinking, and departure from the idea of only one correct solution, encourages and motivates students. When students understand that they each have a unique contribution and valid voice in the problem solving, they are more likely to become engaged and involved. Even students who have struggled in the past will become empowered to contribute and share their thinking, and less likely to rely on the work of more successful students. Because modeling tasks can utilize a broad range of mathematical abilities, a broader range of students can emerge as capable mathematical learners (Lesh and Lehrer 2003; Lesh, Zawojewski, and Carmona 2003; Weber et al. 2010). Not all modeling problems have several feasible solutions, but the process always presents opportunities for different approaches and diversity of thinking, resulting in greater chances for success for students with diverse backgrounds and experiences. As these students gain confidence in their ability to contribute to the problem solving, they begin to develop a sense of mathematical autonomy.

Mathematical Autonomy

Autonomy is a key component of student motivation. The autonomous student is more persistent in solving problems and demonstrates self-regulation in learning (Deci and Ryan 1987). Autonomous learners express more curiosity and desire greater challenge (Deci et al. 1991; Turner, Warzon, and Christensen 2011). Students are more engaged in learning when they believe they have an important contribution to make and that others consider and respect their ideas. As they build confidence in their own thinking, their intrinsic motivation increases. When students are motivated from within—when they are excited about participating in their own learning for learning's sake rather than because of pressure or external rewards—they become autonomous learners. They are viewed, and view themselves, as developing math experts. They have the confidence to make mathematical decisions, to approach a problem from one direction and, if necessary, change direction and try another. They will use the mathematical tools in their toolkit and decide when and how they might be helpful in a given problem. A vital part of our work as math teachers is to encourage our students to become originators of ideas rather than merely recipients of content (Stefanou et al. 2004).

Our job as teachers is to present the tools, show students when and how they are used, and then provide a context in which they can choose the appropriate tools for a given problem. When we provide a larger toolkit, students begin to approach problems in a variety of (perfectly valid) ways and gain confidence in their abilities. The more students are given opportunities to make mathematical decisions, and the more they are encouraged to use their tools to explore and reason about mathematical problems, the more autonomous they become. As students become more autonomous, they grow increasingly confident in offering their ideas and methods and better able to take the initiative when presented with a novel task.

A student who is mathematically autonomous will:

- Decide which tools and approaches are appropriate to the problem at hand;

- Use a variety of approaches and representations to investigate and solve a problem;

- Decide whether an argument makes sense and is reasonable in the context of the problem;

- Justify the appropriateness of their solution, explain why it makes sense, and convince their group of the reasonableness of their solution;

- Attend to and comment on the work of others;

- Present their work to the class; and

- Decide what further investigation might be interesting or necessary.

(Weber et al. 2010; Yackel and Cobb 1996)

➤ Student Autonomy

When the teacher is excited about learning, interested in students' ideas and thinking, and willing to explore new ways of teaching and learning, he or she will be rewarded with students who are excited about learning and proud to consider themselves mathematical learners. When teachers are open to finding different ways of solving a problem, they are more likely to create students with mathematical autonomy.

In order to create and support autonomous learners, the classroom must be a place that encourages choice and provides positive feedback regarding competence. Students must be assured that they are capable of learning, and that their reasoning and justifications are respected by their peers as well as by the teacher (Deci et al. 1991). In their observations of classrooms that support autonomy versus those that do not, Reeve, Bolt, and Cai (1999) found that those teachers rated high in autonomy support listened to the students more often. In addition, high-support classrooms more often allowed students to handle and make decisions about manipulatives and instructional materials. Teachers in high-support classrooms were more likely to solicit student thinking and less likely to give solutions and specific directions.

A classroom that supports mathematical autonomy has:

- A physical arrangement that encourages student collaboration;

- Problems that encourage student engagement, and provide the opportunity for a variety of approaches;

- Appropriate manipulatives and instructional materials available for student exploration and use;

- Time to explore and try a variety of approaches and representations; and

- Time for students to share work.

A teacher that supports mathematical autonomy will:

- Provide a classroom culture of respect for the thinking and reasoning of others;

- Ask directed questions that encourage student thinking rather than provide solutions or specific direction;

- Spend more time listening to student questions and reasoning, rather than lecturing and directing;

- Assure all students that they are competent, and their ideas are worth sharing with others; and

- Encourage students to be intrinsically motivated, rather than trying to motivate by pressure, deadlines, threats, or rewards.

(Weber et al. 2010; Yackel and Cobb 1996)

When a student believes he or she is truly developing as a math expert; when a teacher is willing to share decision making and provide meaning in lessons; and when a classroom culture supports discovery, investigation, and different approaches and ideas and recognizes the contributions and approaches of all, students are more likely to dive into a problem and less likely to defer automatically to the question, "What do we do now?" As students become autonomous, they will be more likely to identify a problem, discuss methods for approaching the problem, and begin investigating and discussing different possible methods of solving the problem.

► Modeling and Mathematical Autonomy

Good modeling problems provide opportunities for all students to be engaged and to be important contributors—regardless of their prior history of success or failure in math or their preconceived notions that tell them, "I'm just not good at math." When students' opinions, viewpoints, decisions, and ideas are respected and considered, they develop a deeper sense of mathematical autonomy. We all know the students who, when presented with a problem, say, "I don't get it" before they have even read it or thought about it. As students develop mathematical autonomy, they become more confident in their ability to contribute to the discussion and more motivated to attempt a solution.

The introduction of modeling into the math classroom across all grade levels will increase student understanding, interest, and autonomy. Whether engaged in shorter, more focused modeling problems or more extensive, multi-day projects, students gain confidence in their proficiency as mathematical learners and make connections between mathematics concepts and real-life applications. All students, regardless of background or history with mathematics, have the opportunity to contribute and learn through experiences with mathematical modeling.

Rich Mathematical Modeling Tasks

Always the beautiful answer who asks a more beautiful question.

—E.E. CUMMINGS

As teachers make the foray into mathematical modeling in their classrooms, it is important that they use rich tasks. Unfortunately, too few textbooks contain such tasks, so teachers must scramble to find or develop their own. Inevitably, the question arises, "How do we know a good task from a bad one?" In this chapter, we examine the characteristics of a rich modeling task and talk about how to find good tasks as well as how to develop them.

Characteristics of Rich Modeling Tasks

A rich modeling task has several characteristics:

- It is accessible to learners with a wide range of abilities.

- It has some basis in real-life experience.

- It lends itself to a variety of approaches and representations.
- It encourages collaboration and discussion.
- It is interesting and engaging.
- It sparks students' curiosity and promotes decision making.
- It encourages creativity, individuality, and variety in the application of knowledge.
- It provides opportunities for extended learning and challenge for advanced learners.

(Ahmed 1987; Piggott 2011)

Let's take a look at some of the characteristics of a rich task in more detail.

➤ Accessibility to All Learners

The first characteristic should be obvious to almost any teacher in almost any classroom: the students in the desks in front of us come with a wide range of foundational knowledge, mathematics background, and experience with mathematics. If a task is truly rich, it will provide interest, motivation, and challenge for all our learners. It will not encourage our most struggling students to opt out and allow the more proficient students to take over, provide direction, and supply solutions. By the same token, it will not be quickly and easily solved by more advanced math students, but will provide challenge and extensions for these learners as well. A good task will provide opportunities for all learners to contribute with confidence. As we investigate rich modeling problems in this chapter and the rest of the book, we will see how a good task can be accessible to all learners.

➤ Real-Life Tasks

The second criterion for a rich task has been the subject of extensive discussion among teachers, mathematicians, and researchers for many years. What, exactly, is a task with some basis in real life? Some researchers go so far as to use "really real" to distinguish a task from merely "real." Still, there is evidence that *authentic* tasks can result in greater engagement, motivation, and investment in the mathematics content involved (Blumenfeld et al. 1991; Leinwand 2009; Schiefele and Csikszentmihalyi 1995).

But the question remains: What is an authentic task? I would suggest that an authentic task is one with which students have some real-life experience. Dan Meyer (blog.mrmeyer.com) has been engaging in an ongoing discussion that he calls "Real-World Math That Isn't Real to Students." In this discussion, he investigates (and polls math teachers and others about) whether it is more important for a task to be "real world" or interesting, and concludes that, "Real-world math doesn't guarantee interest." The challenge here is to incorporate tasks with scenarios that are interesting to students, even if they may not be "really real."

I have used many problem-solving and modeling tasks that are, admittedly, not "really real," but are, nonetheless, interesting and engaging to students. For instance, mathematics with a literature connection can be very interesting and engaging to students despite the fact that the scenarios investigated would not be called "really real." A resource on the NCTM website called "Real World Math" (NCTM 2012) presents a collection of lessons from the "real world." Some of these present problems students might encounter in their lives outside of school (e.g., "How are basketball teams ranked and bracketed for March Madness?" or "When should I or should I not use a credit card?"), but others have connections to the familiar without being "really real." An example of a lesson with a literature connection, some relationship to students' lives, and high interest for students is "The Legend of Paul Bunyan: An Exploration in Measurement" (Buhl, Oursland, and Finco 2003). In this lesson, students use clues from the story of Paul Bunyan and Babe, the Blue Ox, along with their knowledge of scale models and proportions, to estimate the size of Babe's horns and Paul's frying pan. Some researchers would argue that this is not an authentic task because it is not "really real." Granted, the investigation is based on a legend, there is no Blue Ox, and there is no frying pan with an area of one acre. Students will never, in real life, be asked to find the distance between the horns of an ox, but students are engaged in the story and the accompanying lesson and activity. They can apply proportional reasoning using their own body proportions, knowledge of the basic sizes of animals, and nonstandard measurements to engage in this problem. It is a high-interest problem that can later connect to real-life applications. Students will gain an interesting application of prediction, proportions, and scale models that they may later use to find, for instance, the true distance between two points on a map or the dimensions of a room from a blueprint.

Beginning modeling tasks might ask students to find patterns in constructing a tower with blocks or tiles (see the Block Tower problem in Chapter 3) or to determine how many pebbles might bring the water in a pitcher to drinking level for the crow in Aesop's fable (see The Crow and the Pitcher problem later in this chapter). Obviously, students will not be building towers in real life, nor will they be required to drop rocks into a pitcher to get a drink, but the connections with objects and situations with which they are familiar (e.g.., patterns, a poem or story, water displacement) can make important connections between math concepts and the world around them.

Remember, the objective is to present tasks that are interesting and can help students understand that the concepts they are learning in their math classrooms are not completely separate from their lives outside the classroom. Do not stifle your creativity by imposing false limits regarding what is a "real-world" application. Students will be engaged by problems that spring from literature, from other subject areas, and from their imaginations, as well as from their "really real" lives!

➤ Multiple Approaches and Representations

The third criterion of a rich math task is that it lends itself to a variety of approaches and representations. Our students approach mathematics from very different perspectives, and when we can recognize and respect that fact, we can provide more opportunities for students' success. When a variety of approaches can lead to an appropriate solution, students can become confident in their abilities and learn from one another. I have presented some variation of the following problem at several conferences attended by math teachers from various grade levels:

> James collected some Halloween candy. He gave $\frac{1}{2}$ of his candy to his brother. He gave $\frac{1}{3}$ of the remaining candy to his sister. He gave $\frac{1}{4}$ of the remaining candy to his mother, and kept the rest for himself. What fraction of the original candy did James keep for himself?

Take a moment to solve this problem yourself. Typically, algebra teachers will try to solve the problem algebraically (frequently resulting in frustration). The solution might look something like this:

$$x - \frac{1}{2}x = \frac{1}{2}x \text{ [brother]}$$

$$x - \frac{1}{2}x = \frac{1}{2}x \text{ [remaining]}$$

$$\frac{1}{3}\left(\frac{1}{2}x\right) = \frac{1}{6}x \text{ [sister]}$$

$$x - \frac{1}{2}x - \frac{1}{6}x = \frac{1}{3}x \text{ [remaining]}$$

$$\frac{1}{4}\left(\frac{1}{3}x\right) = \frac{1}{12}x \text{ [mom]}$$

$$x - \frac{1}{2}x - \frac{1}{6}x - \frac{1}{12}x = \frac{3}{12}x$$

So, James got $\frac{3}{12}$ (or $\frac{1}{4}$) of the candy!

The algebra teachers (including me, the first time I did this problem) are usually quite proud of themselves if they get this right, and extremely frustrated if they cannot figure it out.

Some teachers will solve it by guess-and-check—substituting a value for the number of candies collected (with greater or lesser success):

> If James started with 100 pieces of candy, he gave 50 pieces to his brother. That left him with 50 pieces. He gave one-third of that ... oops! That doesn't work. Let's try a number that is divisible by 2, and 3, and 4. Let's start with 24 candies. Brother got 12 pieces, with 12 left over. Sister got $\frac{1}{3}$ of of 12 (4 pieces), with 8 left over. Mom got $\frac{1}{4}$ of 8 (2 pieces), with 6 left over. That left 6 pieces for James. So James got $\frac{6}{24}$, or $\frac{1}{4}$ of the candy!

After a few secondary teachers present their work, I invite an elementary-grade teacher to present hers or his. It typically looks something like this:

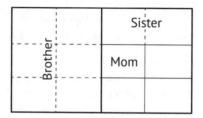

James is left with $\frac{3}{12}$ (or $\frac{1}{4}$) of the candy. This solution usually elicits gasps and giggles from the teachers who tried (often unsuccessfully) to solve the problem in a more complicated manner. It is clear that even we, as teachers, approach problem solving using the techniques, methods, and tools with which we are most comfortable. It is also clear that we can learn from one another when we view the same problem from a variety of approaches, using different representations. Most teachers (and students) realize that the pictorial representation of the Halloween candy is the easiest to do, and the clearest to understand. More important, a comparison of three (or more) methods will produce some interesting and informative insights.

Teachers at conferences and students in classrooms can gain a clearer understanding of the steps of the algebra problem by looking at the picture representation. Students, too, can understand, by comparing the guess-and-check method to the picture, why 100 candies did not work but 24 candies did. The variety of approaches and representations allows access to the problem and enhances comprehension for all learners.

➤ Collaboration and Discussion

The fourth characteristic of a rich modeling task is that it encourages collaboration and discussion. The teaching strategies for modeling problems vary, but in every case should involve some degree of collaboration and discussion. Many tasks (such as the Halloween Candy task) require students to first work independently, with guidance and encouragement from the teacher as needed. When students have reached an independent conclusion, they may share their results, approaches, and representations within collaborative groups and discuss the similarities and differences. Other tasks, such as the three tasks later in this chapter (pps. 24, 25, and 26) lend themselves to small-group collaboration from the start. Students gain new insights and perspectives by hearing from their peers; even struggling students begin to gain confidence in their ability to problem solve when given the opportunity to explain their reasoning on a problem like the Halloween Candy problem or when involved in group discussions on problems like 100 People, Taco Cart, or The Crow and the Pitcher.

➤ Engagement, Curiosity, and Creativity

The next characteristics of a rich task are fairly self-explanatory and go hand in hand with each other as well as with other characteristics. When a task is interesting and engaging, and gives a problem that interests students, they will persevere in the problem solving. A rich task presents students with a variety of possible approaches and representations, allowing access for all learners. When students problem solve, they must first decide which tools and what approach to try. As the problem unfolds, students may run into obstacles and roadblocks that will require them to make more decisions both individually and collaboratively, and, often, will demand creativity and variety in the way they apply their knowledge. In the Halloween Candy task, for instance, students initially decide which approach to use to solve the problem (e.g., algebra, guess-and-check, pictorial). If they hit a roadblock (starting with 100 candies does not work), they need to adjust their thinking and decide how to proceed. They might dig into their toolkits and realize that divisibility and multiples might be helpful, or they might decide to take another approach to the problem. When a group collaborates, teachers may ask them to decide which approach they prefer and to explain it to the class. The rich task sparks curiosity, encourages creativity, and offers multiple opportunities for decision making.

➤ Opportunities for Extension

The final characteristic of a rich modeling task is often the missing piece in problem-solving experiences in the classroom. We can probably all relate to the times we have presented students with an engaging task, only to find that one group finishes quickly while another group struggles with the problem. This can result in one of two scenarios: a frantic attempt to keep the more advanced group busy (or quiet) while the other group continues their work, or an attempt to hurry along the slower-paced group in order to keep everyone on task. A rich task keeps all students engaged by including challenges and extensions for more advanced learners to work out while other students continue to work without the pressure to "hurry up and finish." We will investigate extensions on other problems later, but let's look at an example of a further challenge in the Halloween Candy problem. Teachers can ask students who finish early some additional questions such as, "Give two examples of the number of candies James might have collected on Halloween. Find the number of candies that went to each family member," or "Find the fraction of the *total* number of candies that each family member received." An excellent extension asks student to write their own, similar problem and demonstrate that it works. For the Halloween Candy problem, you might ask them to write another problem giving different fractions of candy to family members and to show their solutions.

Word Problems Versus Modeling Tasks

In order to provide students with rich modeling tasks, we need to understand the distinctions between a word problem, a problem-solving task, and a modeling task. Let's see how a rich modeling task can "grow" from a typical textbook word problem:

We start with a word problem:

> A rectangular garden is 16 feet long and 5 feet wide. Find the perimeter of the garden.

This problem applies a particular formula (Perimeter $= 2L + 2W$) in a real-life context. We typically introduce it after students have had ample practice calculating the perimeter of rectangles without context. Although the word problem does not really address a real-world situation, it does provide a real-life context for the routine exercises students have been practicing.

The problem-solving opportunity goes further:

> What are the possible areas of a rectangular garden with perimeter of 80 feet?

The problem-solving opportunity asks students to engage in some critical thinking in addition to application of the formula. In this problem, students may use guess-and-check or look for patterns as they discover how the areas change, even with a fixed perimeter. Depending on grade level and ability, students may discover a general equation for the area of a rectangle with a perimeter 80 feet [$A = L(40 - L)$].

Now look at the modeling task:

> You buy an 80-foot length of fencing on sale at the garden supply shop with which to enclose a rectangular vegetable garden. You have plenty of room for the garden, and want to enclose the largest area possible to plant your vegetables. Find the dimensions of the rectangle that would enclose the greatest area with the given length of fencing.

This task meets the criteria of a rich modeling task. It combines a real-life scenario with which students might be familiar with an engaging problem that piques interest and curiosity. The solution may be counterintuitive (many students will believe that, with the same amount of fencing, you will enclose the same area) and will provide opportunities for a variety of approaches and representations. The problem is accessible to all learners who have a fundamental understanding of area and perimeter and helps students make important connections between the formulas and real life. The problem lends itself to extensions for advanced learners; for example: (1) Do the same problem, placing one side of the rectangle alongside a garage wall; (2) Would you get the same area with another shaped figure (a triangle, a circle ...)? (3) Can you find a formula you could use to find the maximum area given *any length of fencing*? Teachers will find that students who investigate a modeling problem will develop a much greater depth of understanding—and therefore have a richer *experience*—of the mathematics involved than if they simply solve several area and perimeter exercises or word problems. Exchanging the typical "Do problems 1 through 33" assignment for one rich modeling task will be a valuable tradeoff in terms of student engagement, application of knowledge, and depth of understanding.

Finding Rich Mathematical Modeling Tasks

Textbook publishers are in the throes of developing materials that address the Common Core Standards for Mathematical Practice, but there are several other good resources where we can find rich modeling tasks.

➤ Sources for Existing Tasks

In the digital age, it is possible to find a host of good (and bad!) classroom tasks. I have done quite a bit of legwork (keyboard work?) searching for sources of good modeling tasks and have found that, in the words of problem-solving guru George Polya, "Good problems and mushrooms of certain kinds have something in common; they grow in clusters. Having found one, you should look around; there is a good chance there are some more quite near" (Polya 2008, 65).

MARS Tasks

One excellent resource for rich modeling tasks and performance assessments is the Mathematics Assessment Resource Service (MARS). The MARS tasks offer students grade-appropriate, standards-based problems that spark interest and present challenge to students of varying abilities, and are good resources for teachers. An example of a MARS task for middle school is 100 People.

> There are approximately seven billion (7×10^9) people in the world.
>
> In the 1990s researchers calculated that if there were just 100 people in the world there would be
>
> - 20 children
>
> - 25 people would not have food and shelter
>
> - 17 people would speak Chinese
>
> - 8 would speak English
>
> What *fraction* of people in the world do not have food and shelter?
>
> How many people in the world do not have food and shelter?
>
> In a world of 100 people, how many would live in the United States?
>
> (Courtesy of the Mathematics Assessment Project: www.map.mathshell.org.
> Copyright © 2011 Mathematics Assessment Resource Service.)

This task meets the criteria of a rich modeling task. It is a high-interest task that all students can relate to. The task addresses many content areas including fractions, scientific notation, number sense, scales, and proportions. It requires proportional reasoning, but students may use a variety of approaches. The problem is accessible to all students, whether or not they have a clear understanding of the traditional method for solving proportions using cross-multiplication, or their understanding of the academic language terms "cross-products," or "means and extremes." In fact, it can provide important foundational knowledge of proportional reasoning that could lead to a discussion of traditional ways to solve proportions. It also lends itself to extensions

Modeling with Mathematics

for advanced students ("Find the world population today and the United States population today. How accurate were your estimates?" or "Find another world statistic—the number of people in India, the number of people who are college educated, the number of people who are Buddhist— and find the associated number in the world of 100 people.").

The MARS tasks address a variety of standards, and include rubrics as well as teacher information and suggestions. They provide a rich mathematical experience for students, and are easily accessible to teachers (http://map .mathshell.org/materials/tasks.php).

Three-Act Math

Another resource for high-interest, rich mathematical modeling tasks is 3-Act Math, originally developed by Dan Meyer. These tasks typically begin with Act 1—a visual (often video) introduction to the challenge. This act is the "hook" to the problem: it requires little of the student, yet piques interest in the task by presenting an interesting problem. In Act 2, students are given some of the information they might need to solve the problem. It is the job of the student (with the teacher's help) to dig into the toolkit to figure out what tools and skills might be needed to solve the problem. Act 3 provides the resolution to the conflict. In this part of the problem, students see for themselves whether their expectations agree with the solutions given (Meyer 2011). The 3-Act Math tasks vary by grade level and topic, and are easily accessible online (for a spreadsheet of all of the tasks, see https://docs.google.com/spreadsheet/ccc?key=0AjIqyKM 9d7ZYdEhtR3BJMmdBWnM2YWxWYVM1UWowTEE#gid=0).

Taco Cart is an example of a 3-Act Math problem (http://threeacts .mrmeyer.com/tacocart/). In Act 1, students are shown an overhead video with a dilemma: how to most quickly reach the taco cart across the beach. Dan predicts he would reach the cart more quickly by walking partway in the sand, then making a turn and walking the rest of the way on the road. Ben thinks it would be faster to walk in a direct line across the beach to the taco cart. Students are asked to predict which man is correct. In Act 2, students see a visual of the selected paths toward the taco cart with their respective dimensions as well as the men's speeds. At this point, students use their knowledge of rates, times, and distances to determine which man will reach the taco cart first. In Act 3, students are shown the entire video, showing each man's walk to the taco cart. This task includes several "Sequels" (with related videos)—extensions of the problem that ask students to find the location of the cart that would have both men reaching it at the same time, as well as asking students to determine the path that would take the least amount of time. The task provides an engaging, challenging problem that can be solved using a variety of methods by learners of varying abilities. It provides an extended challenge for more advanced learners and provides teachers an opportunity to discuss the problem in light of real-life applications of rates and geometry.

The 3-Act Math tasks offer high-interest lessons that engage students, promote discussion, and increase understanding. I have used many of these tasks with my students, and am always gratified at the amount of math talk that goes on as the students discuss the problem and the possible methods for approaching the problem. Be sure to consider tasks that are not necessarily targeted to your grade level. Many tasks aimed at higher grades are interesting and accessible to lower grades, and vice versa. I love the Bucky the Badger task (http://mrmeyer.com/threeacts/buckythebadger/) and have used it with middle school and high school classes at the beginning of the year or after a long vacation. In this task, a short video clip (featuring Rachel Maddow) explains that every time Wisconsin's football team scores, their mascot, Bucky the Badger, has to do push-ups equivalent to Wisconsin's total score. The first act reveals a game where Wisconsin scored 83 points! Act 2 asks students to predict the total number of push-ups Bucky had to do throughout the game, and brings up the question of whether the order of point-scoring matters (touchdowns and then field goals, or vice versa). Students then work on possible combinations of touchdowns and field goals totaling 83 points, and calculate the total number of push-ups. Act 3 answers the push-up question, and also questions whether the same "Bucky" is doing the push-ups at the end of the game. Although it addresses third-grade standards, I have used it successfully to get middle and high school students excited about math and thinking mathematically. The task lends itself to powerful math talk where students use strategies to justify their reasoning ("We can check whether order matters by calculating push-ups for a touchdown and then a field goal, and then calculate a field goal first and a touchdown second"). There is always a big, vocal payoff when the students respond to Act 3 (the solution) with sighs and cheers!

NCTM Illuminations

Another resource for good mathematical modeling tasks is NCTM's Illuminations website (http://illuminations.nctm.org/). Many of the lessons include rich modeling opportunities. A few years ago, I participated in the Illuminations Writing Group and developed two lessons for publication on the site. I share one of them with you and encourage you to look it up in its entirety on the website.

In an attempt to provide students with a real-life experience of linear functions and slopes using a literature connection, I developed a lesson around the Aesop fable "The Crow and the Pitcher" (Wolf 2013a). In the story, the thirsty crow tries to reach the water at the bottom of a pitcher by dropping pebbles into the pitcher in order to raise the water to drinking level. In this task, students simulate the situation using a graduated cylinder (the pitcher) and marbles (the pebbles) to determine the rate of change in the height of the water with each "pebble" added. Students create a physical model as well as a graph of their data to solve the problem and, with directed questioning, come to a clear, practical application of slope, y-intercept, domain, range, and linear functions. The task provides an intriguing and engaging problem, multiple ways to represent the problem, some unexpected results, and an opportunity for participation of all learners. It makes a connection between some rather abstract concepts and a real-life problem and encourages collaboration and decision making. It also provides ample opportunities for extensions for advanced learners and adapts easily to different grade levels and specific standards.

I have used the video of The Crow and the Pitcher lesson from my classroom at conferences and professional development sessions. Frequently, when I present lessons and ideas about mathematical modeling, I get responses like, "Well that's great for you, but it would never work with *my* students." I ask those teachers (and any skeptical readers who might be thinking the same thing) to look carefully at the students in the video. They are clearly engaged in the problem and the mathematics, and are excited about what they are doing and learning. This lesson was conducted in one of my Community Day School classes, where students, for the most part, have been expelled from traditional school. They are troubled, at-risk kids. I get a little choked up every time I watch one of them make a prediction and then exclaim, "Oh yeah!" when his prediction is correct because I know they have had a long history of failure in school and in mathematics. Yet despite their history, these students get engaged and excited when presented with rich tasks. And despite, in many cases, a lack of foundational mathematics background, these students have access to the tasks at hand and have confidence that they can contribute. If these students can do it, so can yours!

➤ Developing Your Own Modeling Tasks

When you become familiar with the characteristics of a modeling task and can recognize a good, mathematically rich problem, you can begin to develop your own problems. There are several approaches you can take when developing your own problems, including beginning with textbook word problems, writing tasks based on students' interests, and connecting mathematics to students' current areas of study in other classes. Whichever approach you take, remember that "Two of the most important attributes of real problems is that they should be meaningful and interesting" (Lesh 1981, 251).

Adapting Textbook Word Problems

Even traditional textbook word problems can provide a good jumping-off point for creating powerful modeling tasks. With practice, you can turn a routine word problem into a richer, more engaging and open-ended modeling task. Let's look at a typical word problem that might be found in a seventh-grade math textbook:

> You have started a new job, and have created a budget. The budget for entertainment is 25 percent of your weekly pay. If you plan a weekly "movie night" that will cost you approximately $15, how much will you need to earn in order to stay within your budget?

This problem could serve as inspiration for a modeling problem like this one:

> You want to get a part-time job to earn some extra spending money. You have figured you need at least $15 per month for entertainment, $15 per month for food, and $20 per month for extra expenses. You can choose between the following jobs:
>
> Babysitting: $6 per hour
>
> Lawn-mowing and yard work: $7.50 per hour
>
> Extra chores around the house: $5 per hour
>
> Filing and office work: $8 per hour
>
> Housecleaning: $9.25 per hour
>
> Choose a part-time job (or a combination of jobs) that you would like to try. Figure out how many hours per month you would need to work in order to pay for your expenses. Write up a budget, determining what percent of your income you will devote to entertainment, food, and extras.

While the word problem provides good practice of a particular content area with a real-life connection, the modeling task expands on the connection by including decision making, a choice for the students, a variety of possibilities related to student interest, several approaches students could take, and the inclusion of several math concepts in one problem. The modeling task could replace one night's homework, it could be an activity in one class period, or it could be expanded to become a culminating project after a unit on percents. In any case, one word problem can be used to create a modeling task that will provide a rich, thought-provoking, mathematical experience for students.

Read through the word problems in your textbook, thinking about the mathematics involved, the interest level to students, the possibility of making extensions for advanced learners, and the possibility of, perhaps, making the problem more open-ended. Then dive right in and give it a try! As I write, I am flipping through several middle school math texts. Here is a brief "nudge" in the direction of turning a word problem into a modeling task:

CONTENT STANDARD	GENERAL WORD PROBLEM	POSSIBLE MODELING TASK
Operations with fractions	A recipe that calls for $1\frac{1}{2}$ cups of flour makes 2 dozen cookies. Find the amount of flour needed for 5 dozen cookies.	Go through a cookbook and find a recipe you like with at least 5 ingredients. Rewrite the recipe to feed 10 people, and to feed the whole class. What other factors must you consider besides amount of each ingredient?
Linear equations and functions	A pizza restaurant charges $2 delivery and $15 for a large pizza. Find the cost of having 3 large pizzas delivered.	Using delivery menus from several restaurants, determine the number and size of pizzas needed to feed the class. Are there any other options (number of toppings, coupons ...)? Find the cost per pizza and the delivery cost. Decide on the best buy, if all of the restaurants have equal quality pizza.
Statistics: mean, median, and mode	Luz scored 82, 75, 88, and 83 on her quizzes. What is her median score?	Give students the scores from your last class test or quiz (in *another* class period!). Ask them to find the mean, median, and mode score, and which measure is the best indication of the class "average," and why. Have them take the scores of a particular student (e.g., Student 5) and determine what score he or she would need to get on the next test/quiz in order to get a certain average (80 percent? 72?)
Volume of prisms and cylinders	Find the volume of a can that is 6 inches high with a diameter of 3 inches.	Choose a box of cereal (or sugar) and determine which of a set of cylinders will best hold the contents of the box. Test your conclusions.

These are but a few suggestions that may point you in the direction of creating your own modeling problems using your textbook word problems. Just focus on finding a topic with which students might have some real-world experience, and one in which you can create a problem they might be curious about. You will never find a topic every student is excited about, but if the students can make real-world connections with the mathematics, they will certainly be more engaged and motivated to succeed. If you can take the task a step further by making it an *experience* of the mathematics (using local pizza menus, using the students' own scores, actually pouring cereal into a canister), so much the better!

Using Students' Interests

Another way in which teachers can create rich modeling tasks is to develop problems based on students' areas of interest. As you get to know your students, you can begin to incorporate these interests into your math classroom. Find out what they are interested in and how they spend their non–school time. Talk to them and listen to them. What are they talking about? What television shows and movies are they watching? What sports do they play or follow? All of these areas can provide fertile ground for developing math tasks.

Many of my students skateboard, and there are several skateboard parks in our surrounding neighborhoods, so I decided to incorporate skateboarding into the math classroom. I did a little research on types of skateboard ramps, and came up with the following Kicker Ramp problem:

Your brother wants to build 3 different kicker ramps for practicing tricks on his skateboard. The first ramp is 6 feet long, 4 feet wide, and 1.5 feet high. Find the steepness (slope) of this ramp, and then sketch the plans for two other ramps using boards (a) 8 feet long and 4 feet wide, and (b) 12 feet long and 6 feet wide. Decide on the height of the ramp for each and determine the steepness (slope) of all three ramps. How high do you think you could build each ramp in order to make a reasonably steep kicker ramp? Explain your reasoning.

In this problem, students become interested in a real-life situation with which they are somewhat familiar (even if they are not skateboarders). They get a real-life experience with the concept of slope, and can use real-life experience to determine the reasonableness of their solutions (even a nonskateboarder will recognize that a too-steep or too-flat ramp will not be useful). Students may draw or build scale models and represent slopes in different ways (fractions, decimals, percents). The sometimes abstract mathematical concepts will come alive for students when they can connect them to real-life experiences.

It's easy to transform students' many real-life interests into rich modeling tasks. Any sport can provide data for statistical interpretation (easily accessible on websites like www.nba.com, www.nfl.com, or www.mlb.com). Television ratings are perfect for percent of change problems and proportions (www.nielsen.com/us/en/top10s.html). The school site offers opportunities to investigate geometry and slopes (are the stairs or ramps at our school compliant with building codes?). Or use art and architecture to examine proportions and the Golden Ratio. Students are always interested in problems involving (their) money, so capitalize (excuse the pun) on this interest when studying decimals, fractions, percents, and rate conversions. Travel can present an exciting and exotic way to look at rate, time, and distance problems. With the popularity of skateboarding and snowboarding, many geometry concepts have become much more accessible to students (most students will know what a "360" is).

You can even bring your own interests and experiences into your classroom in an engaging and mathematically valid way. I have had a fascination with Mt. Everest and its climbers for many years, and have read everything I could get my hands on regarding this topic. I have shared this fascination with my students and developed math tasks based on the topic. Over the years, my students have studied climbing rates, blood oxygen levels of climbers, and the cost of an expedition, among other Everest topics. My passion for the subject has motivated students to learn more about the mathematics and sparked an interest in the world around them. In *A Mathematician's Lament*, Paul Lockhart reminds us that "Your teaching should flow naturally from your *own* experience in the jungle, not from some fake tourist version with a car on tracks and the windows rolled up!" (2009, 139). The world around us is filled with mathematical opportunity. Use it to show students the beauty as well as the practicality of math. Apply it to your students' interests and your own, to show students that math is real and exciting and important!

Incorporating Students' Other Areas of Study

Another way to create engaging, rich modeling tasks is to incorporate students' areas of study in other classes. A middle school science standard involving natural hazards and catastrophic events easily pairs with a mathematical task comparing the relative intensity of earthquakes. Many resources provide math activities relate to literature, and it is not difficult to adapt them to your students' reading list. For instance, if students are reading *Gulliver's Travels*, a story featuring little people and giants, students could use proportions and scale models to estimate the height of a giant based on his footprint. For this activity, you can produce an enlarged "footprint" and ask students, based on the footprint and their own foot size, to estimate the height of the giant. An eighth-grade social science standard might involve basic facts about the structure of the U.S. government. Students can find the number of their own state's representatives based on state population, and then predict the number of representatives from other states, based on their populations. Physical education classes provide opportunities to compare resting heart rates to postactivity heart rates, or to graph improvements over time in field-goal percentages or time to run the mile. When students understand that mathematics is connected to all areas of their schooling and their lives, they will be more motivated to learn.

In their history class recently, my students had been studying World War II. They were discussing it in class and asked if I had been to Pearl Harbor. In talking about the war, I began to investigate topics that might connect their history lessons and our math studies related to algebraic reasoning. I developed a modeling task based on the story *Sadako and the Thousand Paper Cranes* by Eleanor Coerr (1977). Several students had heard of the nonfiction story about a child living in Hiroshima, Japan, when the atomic bomb was dropped in 1945, and her attempt to fold 1,000 origami cranes before her death from leukemia. The act of origami is, in itself, a rich mathematical experience, and the story makes connections to middle school World History standards, so we expanded on the origami and the historical connections with a mathematical modeling task. Students were curious about how long it might take to make 1,000 paper cranes, and a modeling problem was born! The basic problem is outlined here:

The school wants to display 1,000 paper cranes for Open House. Our class has been asked by the principal to present a plan for displaying the 1,000 cranes, the estimated time it will take to make that display, and the best location and arrangement for the display. Based on the presentations and data, school officials will decide what group will be selected for the job.

In your groups, you will investigate the questions regarding length of the string(s) of cranes and the time needed to create the 1,000 paper crane display. Then you will create a presentation that shares your results and will convince the school officials that your results are correct, and that your group should be selected to create the cranes.

This problem provided us with a rich mathematical modeling experience combining cross-curricular and real-world connections, which we used repeatedly in a variety of mathematical contexts.

As you work with problems like the MARS tasks, the 3-Act Math tasks, and the Illuminations problems, it will become easier to recognize a good mathematical modeling task and develop your own rich mathematical experiences for your students, and you will begin to see the return on your investment of time and energy in the enthusiasm, motivation, and depth of understanding of your students.

Identifying and Investigating Problems

3

Think left and think right and think low and think high. Oh, the things you can think up if only you try.

—Dr. Seuss

Most modeling experts agree that the first step in the modeling process is probably the most difficult and the most important. This step involves identifying the problem and investigating what information is necessary and unnecessary and is the step where students will begin to make decisions regarding the approach they might take and the materials they might need. When teachers begin incorporating nonroutine modeling tasks into the classroom, most agree that this step is the most challenging! In the traditional math class, students do not have to make many decisions. They have been doing a page or two of problems using a particular procedure or algorithm, and then do one or two word problems using the same procedure. Modeling is challenging because we are now asking our students to take responsibility for how they will investigate and solve the problem.

The first step in investigating a modeling problem is to understand it. The student should be able not only to restate the problem but also to identify the necessary (and unnecessary) parts of the problem. When the student under-stands the problem, he or she must next identify what tools and skills will be needed to solve the problem. Finally, students may begin to devise a plan for solving the problem.

Getting Started: Student Autonomy

Any teacher who has been in a traditional math classroom has experienced something like this: We give some background information, hand out the task, and ask students to read and discuss it. Within 5.2 seconds (certainly less time than it takes to read the task), at least one student will say, "I don't get it. What are we supposed to do?" Students in the traditional math classroom have become accustomed to practicing endless exercises surrounding a given concept, and then, perhaps, solving a couple of word problems using the same algorithm or strategy. A rich modeling problem will present these learners with some challenges: Which strategy are we using here? What do we do first? What are we trying to find? When students are used to step-by-step instructions and specific rules to follow, they have a difficult time stepping out on their own. As we help students transition from the traditional math classroom, where they are *learning* math, to a richer environment where they spend more time *doing* math, we are encouraging them to develop mathematical autonomy.

Too many math teachers are used to a classroom where there is one right answer and one way of doing the problems. When we begin to investigate modeling activities and nonroutine problems, we begin to realize that there are many approaches to the same problem and, often, many right answers. Incorporating modeling into the classroom challenges the teacher to really understand the problem, to encourage and support students as they clarify their justifications and explanations, and to also be a thinker and problem solver when interpreting and assessing student work and justifications.

In my years of designing and implementing problem-solving opportunities and modeling activities, I am continually astounded at the variety and creativity of students' (correct) responses to nonroutine problems. When I began this jour-ney, I would view a problem as having one obvious solution, but then the students would explain their reasoning, and I would learn as much as they would! When I was open to a variety of methods and approaches to a problem and worked hard to understand what the students were trying to tell me, I often discovered new ways of looking at problems. I have some problems I use frequently with groups of students. Even after a decade of use, each time I present them I add to my own knowledge and understanding through students' novel approaches.

➤ Starting Simple: The Block Tower Problem

A baby step toward autonomy in a real-life modeling problem can be seen in a pattern activity such as the Block Tower problem. In this problem, students are given the following visual and are asked to re-create it with color tiles and to identify patterns.

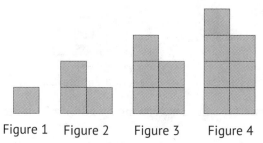

Figure 1   Figure 2   Figure 3   Figure 4

Problems like these can demonstrate to students that there are many ways to view and approach the same problem, and that all students can have an important contribution to make that can add to other students' understanding.

I was recently working with a group of students in a middle school special education classroom. These students had a long history of struggle and failure in mathematics, had spent years in "remedial" courses that consisted of rote learning and calculations, and had little confidence in their own abilities. When first presented with the Block Tower problem, many students balked. This was an "intervention" class where the students spent (by the teacher's description) most of their time working with flash cards, memorizing multiplication charts and fact families, and practicing two- and three-digit multiplication and long division. The students had little experience with problem solving and nonroutine investigations and even less confidence in their own abilities. While some students started constructing figures with the color tiles provided, most sat quietly and waited for instructions. As I went from group to group, I encouraged them to construct the figures with the tiles, and then draw them on their papers. With this prompting, the students got to work.

Start with Exploration: Finding the Pattern

When most students had drawn the first several figures, I asked the class to describe the patterns they found. Initially, only one or two students were brave enough to explain the patterns they saw, but when it became apparent there was more than one way to view this problem, and that all ideas were respected and considered, the room exploded with suggestions. When students understood that there were many possible patterns (and then discovered even more when they created tables and pictures), they became confident to try to find and test their own patterns. This activity became a very rich mathematical environment where students felt free to explore new ideas, make mistakes and correct them, and share their discoveries with others.

Some of the students' approaches are given here:

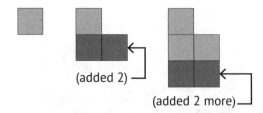

This student viewed the figures growing by adding 2 tiles to the *bottom* of each figure.

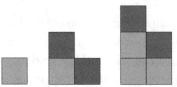

This student suggested growing the figure by adding 1 tile to the *top* of the left column, and 1 tile to the top of the right column on each figure.

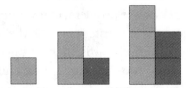

This student constructed the growing pattern by making one column with the same number of tiles as the figure number, and one column with one less than that number of tiles (i.e., one column has 3 tiles, and one column has 2 tiles).

Another student viewed the growing pattern as a rectangle consisting of the figure number times 2, and then removing the upper right tile each time. That is, for the fourth figure, he constructed a rectangle that was 4 × 2, and then removed the upper right tile:

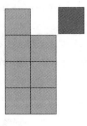

The Block Tower investigation encouraged students to look for new patterns and investigate their own ideas. Even the most reluctant learners became engaged in the investigation when they felt assured that there were no wrong answers. Even when they came up with patterns that did not prove to be true in all circumstances, they were encouraged to use their knowledge to try again.

When students are unaccustomed to nonroutine problem-solving opportunities, they may have a difficult time with the initial investigation and identification of the problem situation in an extended modeling problem. In this case, it may be advantageous to begin with smaller activities such as the Block Tower problem in order to encourage student autonomy, promote motivation, and give students experience with problems for which there are many approaches and viewpoints. In this way, they will be more prepared to approach more complicated problems with a degree of autonomy.

Getting Started: The Teacher's Role

As students begin investigating a nonroutine modeling problem, the teacher's role as facilitator is vital. He or she must pose guiding questions, rather than answer every question students ask. George Polya (2008) was an extremely influential teacher and mathematician in the twentieth century. His seminal work, *How to Solve It*, was first published in 1945, but its clear instruction in teaching and learning problem solving is still relevant today. He directs teachers to "try to understand what is going on in the student's mind, and ask a question or indicate a step that *could have occurred to the student himself*" (1).

When beginning to investigate and identify the problem in a modeling task, there is a delicate balance teachers must strike between giving students direct instructions and answers and allowing them to flail around frustrated and without direction. Nicole Rigelman (2007) compares two teachers' approaches and discourse with students regarding a pattern problem similar to the Block Tower problem. Teacher A directed students to the step-by-step problem-solving poster on the wall and asked them to find the number of tiles in the 25th figure. She repeatedly pointed students to the strategy that would provide the one correct answer to the problem, without any need for individual student thinking or depth of understanding. Teacher B asked students questions such as, "What do you see in the pattern?" "Would you show us what you see?" and "Did anyone see the problem another way?" (310). While the first teacher is giving the students strategies for problem solving, there is little room for critical thinking and flexibility in the approach. The second teacher spends time and energy investigating student reasoning and sense making and considering others' ideas and approaches. The second teacher is providing a deeper, richer mathematical experience for all students, while still providing some direction and support for student thinking. Teacher B empowers students to explore their ideas and to make sense of and build on the reasoning of others.

➤ Questioning

It is particularly important in the initial steps of problem identification for teachers to use questioning techniques and allow wait time in order for students to develop into autonomous learners. It is often tempting to give students specific instructions or direction when they expresses helplessness, but we must remember we are not just solving this one problem—we are trying to create critical thinkers who are capable of taking the initiative when presented with real problems. We can ask directed questions that will point students in the right direction without taking away the power they gain from decision making, risk taking, and confidence in their own ability to solve problems. Even when students are "stuck" at the very beginning, good questioning techniques can propel them forward and facilitate identification and investigation of the problem at hand.

Students accustomed to a traditional approach in their math classes may be disinclined to approach a problem without explicit instructions. When students ask, "What are we supposed to do?" the teacher might ask them some questions in return:

- Have you read the instructions?
- Tell me what the problem is about. What are you supposed to find out?
- Can you restate the problem?
- Can you make a sketch of the problem?
- Can you represent the problem using manipulatives?
- Have you seen another problem like this one?

After students have been given time and opportunity to reconsider the problem, ask students to restate the problem and describe the task.

➤ Helping Students Get Started

In the initial stages of the Block Tower problem, students who have not had experience with problems of this sort may not know where to begin. When asked some of the questions above, the student responses in one class showed an increased ability to move forward with the investigation:

Student: I don't get it.

Teacher: Read the problem again. Set up the problem on your desk using the color tiles and think about what the problem is asking you to find. I'll be back.

The teacher walks away and allows the student to think about the problem on his own, which reduces pressure on the student and also reduces the temptation for the teacher to step in with direct instructions and answers. After a short period of time, the teacher returns to the student:

Teacher: What have you discovered?

Student: I made the pattern with the tiles. The problem wants to know about the 25th figure.

Teacher: Good. Can you make the next 2 figures? Then maybe you can find a pattern that will help you.

Again, the teacher will give the student time to work and investigate independently. When she returns to the student, the student has made the next 2 figures:

Teacher: Tell me what you've found. How did you make the next figures?

Student: I added 2 blocks to the bottom every time (demonstrates).

Teacher: So how many blocks did you add to the bottom of Figure 2?

Student: Two.

Teacher: And how many on the bottom of Figure 3?

Student: Two more. Four.

Teacher: That's right. Use that information to see if you can find a pattern that will help you find the number of tiles in Figure 25.

Through the questioning and discourse, the teacher is guiding this reluctant learner by first listening. When the teacher has an understanding of the way the student views the problem, she can provide direction through questions that relate to the student's thinking. The teacher articulates a step that "could have occurred to the student himself" (Polya 2008, 1), giving the student a direction for the problem based on the student's own reasoning. The teacher prompts the student to think about the problem and make a physical model, and then, through questioning, the student is able to articulate his thinking and reasoning. Through more guided questioning and verbal cues, the student moves forward in the problem. Because the teacher has provided wait time for student thinking, and because the teacher's questions are based on the student's own ideas, the student gains confidence in his reasoning and his ability to progress.

Identifying More Problems: Old MacDonald's Farm

A classic algebraic-thinking, modeling task goes something like this:

> Old MacDonald has a farm. On his farm, he has goats and chickens. He has a total of 60 animals (goats and chickens), and those animals have a total of 156 legs. If no animal is missing a leg, how many goats and how many chickens does Old MacDonald have on his farm?

Problem identification on this task requires students to understand that they must consider two sets of constraints: the number of animals and the number of legs. When presented with this problem, many students will consider only the number of legs. Depending on their background knowledge, the students will probably display a variety of initial approaches to this problem. Whether using a guess-and-check approach, making a table, or using some other method, many students will quickly come up with a solution. They may report, for instance, that 30 chickens and 24 goats have 156 legs (correct), but fail to consider the other part of the problem: Old MacDonald has 60 animals. (Note to teachers: I made an important discovery with this problem. If you do not teach in a rural area, do not assume your students know how many legs are on a goat or a chicken!)

With guidance and good questioning techniques, teachers can help students move forward. When their excitement over solving the problem turns to frustration, teachers can facilitate perseverance as students revisit the essential problem and understand they have not truly understood nor identified the entire task. Dialogues like the following can get students back on track:

Teacher: Good start, but does your solution answer all parts of the task?

Student: We don't have the right number of animals.

Teacher: Do you have too many, or too few?

Student: We only have 54 animals, not 60.

Teacher: Think about what that means. Do you need more goats or more chickens? Keep working.

A little directed questioning and encouragement can diminish student frustration and enable students to identify all parts of the essential question.

Having done that, they can then move on to the next steps of modeling. With teacher support, they are not left to flounder when they make a mistake but are guided to use their discoveries, and even their errors, to help them solve the task at hand, and move forward with their model toward a solution.

Getting Started: 1,000 Paper Cranes

Let's look again at the 1,000 Cranes problem. In the initial stage of the task—investigation and problem identification—the students develop a plan for making a string of 1,000 paper cranes for Open House. They consider the given information and decide what is necessary and unnecessary. They make decisions about materials and math tools needed to solve the problem and access prior knowledge about problems like these. For this problem, students identify the unknowns: length of a string of cranes, time to make cranes, design ideas, spacing of cranes, materials needed, and so on.

➤ In the Classroom

In one seventh-grade classroom, the teacher reviewed the background of the problem (the historical context, the art of origami, the symbolic meaning of the crane) and distributed an activity sheet. Students worked in groups of three or four and were given time to read and discuss the problem. They were prompted to identify the problem(s) they were asked to solve, identify materials they might need to solve the problem(s), and devise a plan for data collection, problem solving and presentation. This class had engaged in earlier problem-solving opportunities and was fairly comfortable with the expectations of the investigation part of the task. One group's discussion went like this:

> **Student A:** I think we should hang the string in the cafeteria.
>
> **Student B:** There's no way it will go across the cafeteria. We should hang it in the office.
>
> **Student C:** Wait a second. We have a lot of work to do before we can decide where to hang it! Let's write some stuff down. First we have to decide how long the string will be.

Student C is taking a leadership role, and directing students back to the real task at hand. He is, essentially, asking the first question of problem solving: **What is the unknown?**

Student B (incredulously): Do we really have to first make 1,000 cranes?

Student D (returning to the activity sheet): Let's write down the problems we're investigating. We need to know how long the string will be. We need to also know how long it will take to make the cranes.

Student D refers the group back to the questions on the activity sheet, and defines again what the task is asking for.

Student C: We're not gonna really make 1,000 cranes. We're just gonna make a few, and figure out how long it would take to make more.

Student C has quickly identified one problem-solving strategy that has been used before in class: **make a big problem into a smaller, more manageable problem.**

Another group struggles with the same problem:

Student E: It'll take us, like, DAYS to make 1,000 cranes. (To teacher) Are we really making 1,000 cranes?

Teacher: Does that seem like a good idea? What if the office staff doesn't select your group to make the cranes? Is that a wise use of your time? What questions are you trying to answer?

Student E: How can we figure out how long the string is if we don't make it?

Teacher: Can you think of any other problems we've done like this one? What strategies did we use?

Student F: We did the ice cream cone problem.

Teacher: That sounds like a good one. Did we measure all 18 scoops?

Student F: No. We measured 3 or 4 scoops, and then figured out the rest.

Teacher: Can you use that strategy on this problem?

Student E: Yeah. We could make a string of, like, 5 cranes and measure it. Then we could find a way to figure out 1,000.

Teacher: Sounds like a good strategy! Easier than making all 1,000 cranes, huh? (Laughs) Write that down on your planning section.

This discussion reinforces one problem-solving strategy (making a big problem into a smaller, related problem), and also introduces another good strategy: **asking students if they recall another, similar problem, and how they solved it.**

A third group is struggling with autonomy. As the teacher nears the group, someone asks:

Student G: What are we supposed to be doing?

Teacher: Tell me what you have read. What are the problems you're trying to figure out?

Student H: We're going to make a string of 1,000 cranes, but we don't have the paper. What are we supposed to do?

Teacher: Let's look at the activity sheet again. What do we need to do before we start making cranes?

Student G (reading): "Investigate the questions regarding length of the string and the time needed."

Teacher: That's right. Can you write down the two questions you're investigating?

Student F: How long is the string and how long will it take?

Teacher: That sounds right. (Students write down questions.) Now think about what you will need and what you will need to do to answer those questions.

Student G: We need the paper and the string. Then we can make the cranes.

Student H: We're not going to make all the cranes, are we?

Teacher: What is the unknown we're investigating first?

Student F: How long will the string be with 1,000 cranes.

Teacher: Have we done a problem like this one before?

Student G: We made the cranes before. (Students had experience with origami and paper cranes in art class.)

Teacher: What is the unknown? Did we do a problem with a similar unknown?

Student H (with uncertainty): How tall is the ice cream?

Teacher: Good. How is that problem the same?

Student H: We figured out how tall, and this one is how long.

Teacher: That's right. I want you to talk about the strategies we used to solve the Eighteen Flavors [Bay-Williams and Martinie 2004] problem and see which ones might work on this problem. I'll be back.

The teacher has used guided questioning to help students identify the problem and begin thinking about strategies to solve the problem. In the 1,000 Cranes task, the problem of making all 1,000 cranes immediately draws the students in, so the transition to an easier, smaller problem is generally a smooth one.

Another group has begun considering the amount of time needed to construct the cranes, rather than the length of the string, but has discovered that the problem-solving strategies are very similar.

Teacher: What have you identified as the problems?

Student I: How long are the 1,000 cranes and how long will it take to make them?

Teacher: It looks like you already have a plan for answering those questions. What have you decided to do?

Student J: We first thought we'd figure out how long it would take one person to make a crane, but then we decided it would be better to all make cranes.

Teacher: So you're using the strategy of making a big problem into a smaller one? Why do you think it's better for each group member to make a crane?

Student J: It'll be more real because we all make them a little different. Some of us are fast and some are slower.

Teacher: So you'll see how long it makes for each of you to make a crane?

Student K: Actually, we thought we'd just make 10 cranes all together and see how long that takes. I thought that would be better.

Teacher: Sounds like a good plan. Get to work!

The students engaged in the problem, and discussion continued about the unknowns and how the students would accomplish the task. Some interesting variables entered the discussion—how the cranes would be spaced (pushed together tightly, end-to-end, or with spaces between them), how much string would be needed on each end, and other general design ideas. Some disagreements occurred in one group when most students pictured the string of cranes in a horizontal configuration (going across the room), and one student insisted the cranes should be hanging vertically from the ceiling (an idea that never occurred to the teacher!). This student was relying on her prior knowledge when she said, "I've seen this at a wedding, and it looked really pretty." The observation added in the variable of the number of strings of cranes needed, and showed the variety of ideas that can burst forth from an open-ended problem in the hands of creative students.

Conclusion

This class work demonstrates the importance of allowing time for students to investigate the problem, discuss strategies for solving the problem(s), and begin to plan the actual problem solving. The group discussion and directed questioning by the teacher encourage mathematical autonomy. Students' ideas are used to "clarify and refine the monitoring and reflecting process" (Woodward et al. 2012). The suggestions and ideas of all students are solicited and respected. If students require more direction, the teacher offers guidance, then leaves the students to their own devices again to see what they can accomplish. Once students decide what questions they have, the materials needed, and what the plan of action will be, whole-class discussion can clarify questions, reinforce reasoning and ideas, and start students thinking about the next phase of the modeling process: formulating a mathematical model, collecting data, and finding solutions.

It is important for the students and the teacher to get the first steps of mathematical modeling problems right. When students, with the teacher's guidance, understand what the problem is asking—what information is important and what is not—and can formulate a plan to solve the problem, their path to completion of the task is smoother and more direct. Students will do less meandering through the problem-solving process, unsure of how to proceed, if they have identified what the problem is asking and have an idea of what they are trying to find out. With proper guidance, questioning, and redirecting from the teacher, students can flow from the initial problem investigation and identification to formulation of the model and work toward a solution.

4 Formulating the Model, Collecting Data, and Finding a Solution

If you only have a hammer, you tend to see every problem as a nail.

—Abraham Maslow

Once students have worked through the first stage of modeling—investigating, and identifying the problem—they are ready to move on to the second stage—formulating the model, collecting data, and solving the problem. This is where many teachers hit the proverbial wall. This stage requires a change in thinking and behavior for both students and teachers, and is the stage about which there is the most confusion. After visiting many classrooms and speaking to teachers, students, and administrators, it has become apparent to me that many well-meaning teachers are asking students to make the great leap from traditional methods to modeling with little guidance. Teachers often express that they have been advised to merely present a task and step back, letting students work through the problem because "Students learn best through frustration." Those of us who have watched students "learning through frustration" realize that most students, when frustrated, merely shut down. I have spoken to many teachers who have discarded the whole notion of mathematical modeling because they believe their students are incapable of mastering the second stage of formulating a model and solving the problem. Believe me when I say there is ample middle ground between directing students' every step in a math problem and leaving them to flail in the dark. This chapter explores ways of guiding students through the modeling process until they are capable and confident enough to approach a problem with autonomy.

Those of us who are parents know that the process of teaching a child to ride a bike requires a gradual "letting go." The kids start with a tricycle, getting used to the idea of setting off on their own—but never going too far, never without three wheels, and with plenty of supervision. Next, we might switch them over to training wheels, allowing them to experience balancing the bike while ensuring they don't tip over and fall down. Although my daughters are now young adults, I remember the first days without the training wheels, as I ran down the street alongside the bicycle, holding on until I sensed they were balanced. Only then, when they were finally prepared to set off on their own, did I let go and watch my girls ride down the street under their own steam.

So it is with modeling. As students are introduced to the concept of modeling and presented with modeling tasks, we must start slowly, and with plenty of support and supervision. We provide hints and guided questions. We remind them of the different tools they have in their math toolkits, various approaches we have used, and provide opportunities for multiple representations. We start them on tasks that require fewer tools and employ concepts with which they are most comfortable and familiar. We let them experience success and gain confidence on these problems, and then, gradually, require more of them as we provide less specific information and fewer suggestions. If we progress slowly and patiently through this "letting go" process, we will produce autonomous learners who are capable of formulating a model, making decisions regarding problem solving, and incorporating a variety of techniques and representations into their models.

Formulating a Model

When asking students to formulate a model for a given task, we are asking them to use mathematics to represent a problem in the real world. Often, the model is not a completely precise representation of the problem, but the mathematics should provide a solution that is useful. For instance, students might be asked to determine which box (pictured on the next page) would be the best choice to pack particular items. In the most basic case, students would use their knowledge of volume to formulate a model comparing the boxes.

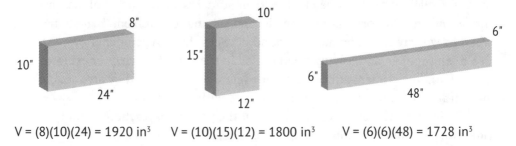

V = (8)(10)(24) = 1920 in³   V = (10)(15)(12) = 1800 in³   V = (6)(6)(48) = 1728 in³

Before performing any calculations, ask students which box they think is best to pack up some "stuff." Students may disagree about which box is "bigger," and some may question what kind of "stuff" we are packing. These are some of the decisions that must be made when formulating a model. Students who have some experience with volume will then begin calculations. Based solely on the volume calculation, the first box is larger than the other two. Students who are accustomed to textbook word problems requiring simple application of a formula will assume the box with the greatest volume must be the best box for packing our "stuff," but they should consider other factors, such as, is volume alone useful enough to solve our problem? If not, what other factors might be considered? We might need to consider what is being packed in the boxes. If we are packing clothing, the configuration of the box will be far less important than if we are packing, say, baseballs. Since the clothing can be folded or smashed to fill the available space, the box with the greatest volume is probably the best. However, if we are packing baseballs (diameter = approximately $2\frac{7}{8}$"), the box with the least volume might accommodate more balls. We might improve the model by considering the thickness of the cardboard in the two boxes. If the cardboard of one is thicker than another, the actual space inside the box will be altered. So, as students formulate a mathematical model, they must make assumptions and decisions that will produce the most useful information for the given problem. They must also give consideration to what information is necessary and unnecessary, depending on the desired goal.

➤ The Ladder of Abstraction

The decision making involved in formulating the mathematical model involves a "delicate shift of attention" (Mason 1989) from the concrete problem at hand to a mathematical representation of that problem by deciding what information is necessary to solve that problem. This process of *abstraction* becomes a powerful tool for students, and is sometimes referred to as the *ladder of abstraction*. The term was popularized by S. I. Hayakawa (1940) in reference to the use of language and is widely used in speech writing, debate, screenwriting, and other language applications, but it is appropriate in the math classroom as well. Dan Meyer (http://blog.mrmeyer.com/2012/loa-abstracting-abstraction/) has a fascinating series on the ideas surrounding the ladder of abstraction in mathematics education and the power of leading students up (and down) that ladder as they consider a problem in math. He synthesizes the concept of abstraction into essentially (1) starting with a question, (2) making predictions, (3) putting the information in context, (4) deciding what information is consequential to the question, (5) naming and representing the aspects of the contexts that are consequential, (6) deciding what to do with our representations, and finally (7) abstracting other situations in the same mathematical context.

In the case of the Packing Box problem, the students could first look at three actual, physical boxes, and predict which one would hold more "stuff." Then, the students need to separate the necessary characteristics of a box from the unnecessary. They are not concerned with the color of the box, or the weight of the box, but depending on the problem, they may need to attend to the thickness of the cardboard or the configuration of the box. They would then use the important characteristics of the box (length, width, height), and decide what to do with them (multiply the dimensions to find volumes, yielding the relatively largest box). Then, when they have calculated volumes of the boxes, they would adjust their model with consideration of the facts of the question (are we packing clothes or baseballs?). When they have experience with the calculations and comparisons of different-size boxes, they can abstract their knowledge of volume in other contexts.

The objective in formulating a model is to find the mathematical representation that provides the most useful solution. Students must decide what information will contribute to understanding the problem at hand and what information is not useful. They must then use the tools in their mental toolkit to determine which strategies would best be used to solve the particular problem.

Using Multiple Representations

The use of multiple representations in the math classroom is not a new idea, but has gained greater importance in light of problem solving and mathematical modeling. *Multiple representations* refers to the use of a variety of methods to describe what is happening in a math problem. The representations might include graphs, pictures, diagrams, tables, equations, symbols, concrete models, and verbal descriptions. When a problem can be represented in a variety of ways, students can begin to understand the varied characteristics of a given problem. They can gain deeper, more flexible knowledge about the problem at hand and formulate a model that is clearer when they can represent a problem in several different ways. Some learners will understand best and thus make progress in problem solving when they can produce and view a visual representation such as a diagram, graph, or picture. Other learners can process information more effectively when presented a numerical representation—a table or equation. Still others may learn best through a verbal description of a problem or process. When a task is viewed from several vantage points, students can get a clearer picture of the problem and how to progress with solving it. They can emphasize or deemphasize different aspects of the problem. Research has shown that using multiple representations in the math classroom increases depth of understanding, student interest, and integration of information in problem solving (De Jong and Van Joolingen 1998; Hiebert and Carpenter 1992; Kaput 1989).

➤ Earliest Stages of Multiple Representations

As we begin the second stage of the modeling process, it is beneficial to make students aware of the various ways in which the same problem might be represented. In the earliest stages (riding the tricycle), it is helpful to start students with some basic problems. Remind your students of the ways in which the problem might be represented—through pictures, diagrams, tables, graphs, equations, and so on—and of the different mathematical tools they might use to solve the problem. For instance, on a basic proportional reasoning problem, students may use several representations and several mathematical approaches. "If 3 pencils cost $.72, what would be the cost of a dozen pencils?" Students might sketch the problem as 4 sets of 3 pencils and then use repeated addition or multiplication to solve. They might create a table showing 3 pencils for $.72, 6 pencils for $1.44, 9 pencils for $2.16, and 12 pencils for $2.88. They might create an equation to find the unit price ($3x = .72$) and then multiply by 12. They might use division to find the unit price ($3\overline{).72}$), or they might set up and solve a formal proportion: $\frac{3}{.72} = \frac{12}{x}$. Remember that traditional word problems typically feature one math concept, and one way of solving the problem, whereas modeling will require students to decide which tool best fits the given problem. As Maslow says, "I suppose it is tempting, if the only tool you have is a hammer, to treat everything as if it were a nail" (1966, 15–16). We want our students to remember they have access to an entire toolkit, and feel comfortable choosing the best tool for the job.

When they first approach a modeling task, some students will respond with the familiar "What do we do?" Encourage them to use whichever representation makes sense to them, and see if they can make progress. Let's look back at the Old MacDonald's Farm problem (from Chapter 3):

> Old MacDonald has a farm. On his farm, he has goats and chickens. He has a total of 60 animals (goats and chickens), and those animals have a total of 156 legs. If no animal is missing a leg, how many goats and how many chickens does Old MacDonald have on his farm?

In one classroom, students were working in cooperative groups of three, had manipulatives (color tiles) available, and had had some experience with problem solving tasks. Still, many balked when presented with this problem and waited for specific instructions from the teacher. The teacher prompted students to "just give it a try" and reminded them of the various problem-solving strategies they had used on earlier problems: "Remember, you can use the manipulatives, draw a picture, make a table, write an equation . . . Just get started, and let's see what you come up with."

Students began working, and after a couple of false starts (a couple of groups quickly came up with answers, such as "30 goats and 18 chickens," forgetting that there had to be 60 animals), students began using a variety of approaches and discussing the problem.

Pictorial Representation
Many students began with a picture or visual representation:

Even though this method may seem cumbersome, many students gain a clear understanding of what they are being asked to do and can begin to reason mathematically once they have made a pictorial representation. In the student work shown in the figure, the group using the picture began a rich discussion and displayed clear mathematical thinking. The teacher asked the students to explain their approach and describe what they had discovered.

Teacher: Tell me about your picture. How did you decide on 30 goats and 30 chickens? How many legs do they have?

Student 1: We just split the 60 in half. The 30 chickens have 60 legs and the 30 goats have 120, so they all have 180 legs.

Teacher: Okay, so what did that tell you?

Student 2: That's too many legs. We have too many goats.

Teacher: So what do you do from here?

Student 1: We started taking away goats. We're not done.

After a period of time, the teacher returned to the group and, seeing they had solved the problem, asked them to describe their reasoning.

Teacher: It looks like you came up with an answer. How do you know it's correct?

Student 3: We got 18 goats and 42 chickens. That's 60 animals.

Student 2: And 18 goats have 18 times 4 legs. That's 72. The chickens have 84 legs. So that makes 156 legs.

Teacher: Show me how you got that answer on your picture. When I was here last, you were pretty far away from 18 goats.

Student 1: We started taking away a goat and adding a chicken and adding it all up, but then we figured when we took away a goat and added a chicken we were just taking away 2 legs every time. Since we needed to take away 24 legs, that's 12 times.

Teacher: Twelve times?

Student 2: We took away 12 goats and added 12 chickens. That makes 18 goats and 42 chickens.

Tables

Some other students used a table for their work:

The teacher asked this group about their decision making.

Teacher: Tell me about your work.

Student 1: We started just guessing numbers, and then remembered we could make a table.

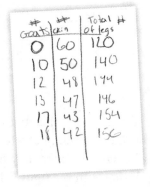

# Goats	# Chickin	Total # of legs
0	60	120
10	50	140
12	48	144
13	47	146
17	43	154
18	42	156

Modeling with Mathematics

Teacher: Tell me about the numbers you chose for your table.

Student 2: The goats and chickens always add up to 60. We counted by tens until we got close to 156 legs.

These students used several different strategies to get to their solution. They started using random guessing and then decided to organize their work into a table (one of the representations the class had been using on a variety of problems). They decided to expedite their work by counting by tens, rather than ones, keeping in mind that the total number of animals was 60. When the group got close to the desired number of legs, they discovered that the solution was somewhere between 40 and 50 chickens, and between 10 and 20 goats. They then used a guess-and-check strategy with deductive reasoning (trying 12 goats, then 13, and then jumping to 17) until they found the correct answer. This group demonstrated understanding of the constraints of the problem and used a logical progression to get to their solution.

Equations

Some of the more advanced students who had some experience with algebra set up the problem using equations. They reasoned that, if there were 60 animals, they could represent the number of one animal (goats) by x, and the other animal (chickens) by $60 - x$. They then multiplied the number of each animal by the number of legs on that animal to get the totals.

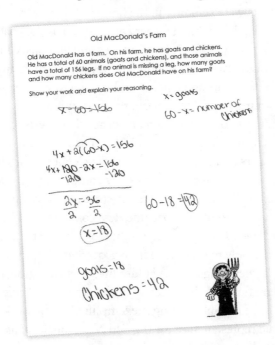

Old MacDonald's Farm

Old MacDonald has a farm. On his farm, he has goats and chickens. He has a total of 60 animals (goats and chickens), and those animals have a total of 156 legs. If no animal is missing a leg, how many goats and how many chickens does Old MacDonald have on his farm?

Show your work and explain your reasoning.

At the conclusion of the task, the groups presented all the different representations, giving students a variety of ways to look at the same problem. Many students said it made more sense to them one way or another and everyone appreciated the fact that there were so many ways to look at the same problem. The pictorial representation made sense to most students, but many of them expressed frustration at having to count the legs repeatedly and commented on the time it took to draw and erase the "goats" and "chickens." Even students who used some reasoning to alter the picture found it difficult to explain their reasoning to the class ("When we take away 1 goat and replace it with a chicken, we take away 2 legs"). Students discussed the use of the table, and many found it to be an organized way to arrive at the solution. One student commented, "We kept forgetting there had to be 60 animals. If we had used that table, we probably would have remembered that."

Although the students in this class had limited experience with representing situations using variables and equations, they expressed interest in the algebraic representation. Many acknowledged that the group who used the equations properly arrived at their answers more quickly and with less work. They realized that algebra could be a useful tool to make their work more efficient.

The Ladder of Abstraction

Let's look at this problem and the multiple representations in terms of the ladder of abstraction. The bottom rung of the ladder would be the concrete view of goats and chickens if students were standing in the barnyard. That view, besides being impractical, would make the problem rather complicated as the goats and chickens move around, making it very difficult to count legs and animals. Climbing another rung of the ladder would involve abstraction: removing or separating unnecessary characteristics of the goats and chickens. This rung might involve using toy goats and chickens, arranging and rearranging and counting legs, removing the element of movement (and disregarding many other characteristics of real goats and chickens, such as sound, smell, color, and size). The next rung would be less concrete and more representational, removing more characteristics from the actual goats and chickens and focusing on the necessary components (the legs). This rung might look like the pictorial example shown. As students move up the ladder, the focus narrows and the representation becomes less concrete (and less connected to actual goats and chickens). The next rung might look like the table of values in the figure above, where the only real connection to goats and chickens is found in the table headings. The higher rungs on the ladder involve abstraction; the relationship between the representation and the concrete objects (goats and chickens) becomes almost unrecognizable. The highest rung is exemplified in the work of the student who used an algebraic representation for the Old MacDonald's Farm problem.

As noted, students begin to understand the power of abstraction when they view the same problem from a variety of representations; many reflected, when reviewing the Old MacDonald's Farm task, that the groups who used equations were able to solve the problem more quickly and easily. When students begin to make connections between the equation $4x + 2y = 156$ and the actual legs on goats and chickens, they begin to understand the power and beauty of abstraction, generalization, and algebraic thinking. It is helpful and informative at this point to deconstruct the equation by climbing back down the ladder of abstraction. Students understand that the $4x$ represents the number of legs on a given number (x) of goats, $2y$ represents the number of legs on a given number (y) of chickens, and 156 represents the total number of legs. The step down the ladder would involve asking students to identify other concrete situations that might be represented by the same abstraction, $4x + 2y = 156$. When I asked a group of students (for whom algebra was a new concept), "What else could be represented by the equation $4x + 2y = 156$?" they first came up with real objects with characteristics similar to the goats and chickens:

"How many cows and ducks have 156 legs?"

"How many horses and people have 156 legs?"

After some discussion and prodding, they came up with:

"There are 156 tires on some cars and motorcycles."

"There are 156 wheels on some 4-wheelers and some bicycles."

Stepping down the rungs of the ladder from the equation to these concrete situations helps students make important connections between the concrete and the abstract, and shows them the power of generalization. After further discussion, some students were able to extend the application of the equation $4x + 2y = 156$ to the following:

"Binders cost $4 and notebooks cost $2, and you spend $156."

"A box has some 4-pound weights and some 2-pound weights, and the box weighs 156 pounds."

Too often, we view the abstract as the final goal. When students have had some experience moving up the ladder of abstraction, we tend to leave them at the highest rung. By moving back down the ladder of abstraction, moving from the abstract to the concrete, students gain a deeper understanding of the process of abstraction and the power of generalization and algebraic thinking. Multiple representations of a problem expose students to a variety of ways of viewing and solving the same problem and facilitate movement up and down the ladder of abstraction.

➤ Guided Multiple Representations

When we are in the "training wheels" stage of mathematical modeling, we might be very specific about requiring students to represent a problem in several ways. Then we ask students to describe what they see and understand in each of the different representations. As students gain confidence with practice using a variety of representations, they will become more autonomous and better able to decide which representations would best serve their needs on a particular problem.

I have used variations on the following problem with students in grades 4 through grade 10. I introduce the Tortoise and the Hare problem by telling the Aesop's fable of the race between the fast, but overconfident hare, and the "slow and steady" tortoise. Then I present students with the following problem:

> The tortoise and the hare leave the Start at the same time. The hare is jogging at a rate of 800 feet per minute, while the tortoise is plodding along at a rate of 200 feet per minute. The hare stops to nap after 5 minutes, takes a 20-minute nap, and starts jogging at the same rate when he wakes up.
>
> Use tables, graphs, pictures, models … to answer the following questions:
>
> > If the racetrack is 5,000 feet long, who will win the race? By how much time? By how many feet?

For students accustomed to working on tasks like this, the prompt may be enough to set them to work, but for students who are unaccustomed to making autonomous decisions and applying tools, this task is overwhelming. To help these students in their earliest experiences with modeling in the classroom, I provide very specific guidelines and representations; for example, I would first ask them to make a sketch or diagram of the problem. For the Tortoise and the Hare problem, it might look something like this:.

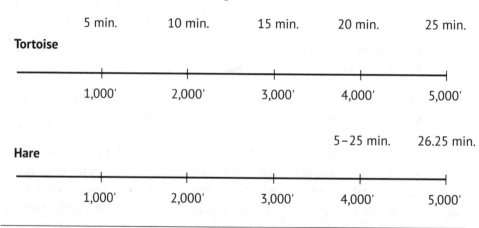

The diagram for this problem may be confusing for students, so I would then ask them to collect data in a specific table:

TIME (MINUTES)	DISTANCE OF TORTOISE (FEET)	DISTANCE OF HARE (FEET)
0		
1		
2		
3		
4		
5		
10		
15		
20		
25		
30		

As students gain more experience with independent work and develop a stronger sense of mathematical autonomy, you can provide them with less direction and specificity. You might set up the table together as a class, then give them a blank table, and, finally, ask them to create a table on their own, deciding on what data to use.

The same guidelines might be used for a graph. With the youngest students, I would provide a completed graph. For older students with little experience, you could ask them to plot points on a graph you have created, labeled, and numbered. Students with more experience and autonomy could create a graph together as a class, discussing appropriate scales on the axes. Finally, when students have ample experience with these tasks, you could give them a blank graph, asking them to decide on the scale and labels.

When the representations are completed, students can analyze the various data in order to make decisions about solving the problem. They may discuss the problems and possibilities of each representation, and decide which they prefer and why.

Sketch

Typically, the visual representation of this problem is confusing to students. It is unclear who is in the lead where.

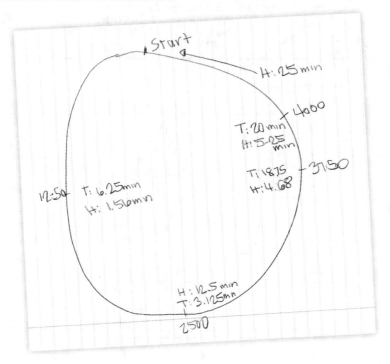

Some students created a picture by drawing two side-by-side tracks, trying to determine where the tortoise and the hare were located after given periods of time. Although they were able to determine that the tortoise won the race by using the picture, this, too, proved confusing. They were still unclear on the questions of "by how much time did he win?" and "by how many feet did he win?" The pictorial representation provided a good lead-in to the use of a table (a higher rung in the ladder of abstraction). The students who started with the tracks quickly added the table to their work, comparing the two as they worked.

Table

Even though the table was further removed from the concrete "race," students could see the relationship between the picture and the table. They acknowledged that the table was an easier method for comparing the distances of the tortoise and the hare at given times.

The students may draw many conclusions by looking at the completed table:

TIME (MINUTES)	DISTANCE OF TORTOISE (FEET)	DISTANCE OF HARE (FEET)
0	0	0
1	200	800
2	400	1,600
3	600	2,400
4	800	3,200
5	1,000	4,000
10	2,000	4,000
15	3,000	4,000
20	4,000	4,000
25	5,000	4,000
30	6,000	8,000

On the table, the students can easily see at what point the tortoise passed the hare (20 minutes / 4,000 feet), and that the tortoise won the race by reaching 5,000 feet before the hare. They can extend the table to find at what point, if the race were longer, the hare would pass the tortoise (between 26 and 27 minutes / between 5,200 and 5,600 feet).

Graph

With a graph, whether created by the students or provided by the teacher, the students can discuss the benefits of this type of visual representation.

They can see clearly the point at which the tortoise passes the hare (the first intersection point), and can see who won the race by determining which line reached 5,000 feet in less time. They can extend the lines to see who would win the race if the distance had been longer, or at what point the hare again passes the tortoise. They can also discuss the meaning of the horizontal line (when the hare is napping, time is still increasing, but distance stays the same). A rich discussion can also follow regarding what the graph would look like if one animal got a head start or a late start. Students may also compare the graph with the table, and discuss the relative benefits and limitations of each.

By comparing the representations, students make important connections and develop a deeper understanding of the various representations. For

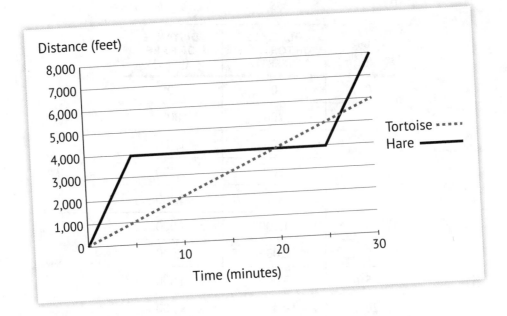

Distance (feet)

Time (minutes)

Tortoise ·····
Hare ——

instance, in the class discussion of the comparison between the graph and the table, some of the conclusions students reached included:

"The tortoise and the hare are both at 4,000 feet after 20 minutes. On the graph, that's where the lines cross. On the table, that's where both distances are 4,000."

"The faster one is the steeper line on the graph, and when the hare stops, the line is flat."

"On the table, the hare is way ahead at the beginning, but the tortoise gets to 5,000 feet first."

"On the table, you can see the tortoise gets to 5,000 in 25 minutes, but the hare gets there between 25 and 30 minutes."

"You can see on the graph, the hare gets to 5,000 feet in about 26 or 27 minutes."

"If the race kept going to 6,000 feet, the hare would win the race."

Many students have become accustomed to creating a table of values merely as a tool to plot points on a graph. When students use multiple representations, they often better recognize that the table itself, even without the accompanying graph, can provide them with a wealth of information. When students can gather information from the table alone, and then in comparison with the graph, their depth of understanding increases. Some students will find the graph easier to interpret, while others find the table easier. In any case, presenting the representations side by side will increase understanding for all students.

Equations

More advanced students may set up a system of linear equations, but as the hare's progress is piecewise, the equations make this a much more complicated representation. It is important to discuss with students the relative benefits of each method, and even if the students are not ready for the piecewise function, you may show them the work. They will understand that the most "advanced" method may not always be the best way to approach a particular problem.

Hare	Tortoise
$f(x) = \begin{cases} 800x & \text{if } x \le 5 \\ 4{,}000 & \text{if } 5 < x \le 25 \\ 800(x-25) + 4{,}000 & \text{if } x > 25 \end{cases}$	$f(x) = 200x$

This same sort of progression can be applied to all sorts of modeling tasks. As students become more comfortable with decision making and autonomy, they will require less and less direction and specific instruction. They should have a clear understanding that they should present the same problem from several vantage points, and will, eventually, be able to do so with relative independence. Still, students will need a nudge to get past roadblocks, and we, the teachers, are always standing by to provide guiding questions to help them progress in the problem solving. When we provide appropriate support and guidelines, rich mathematical tasks become accessible to all learners, and students can progress to the point where we can "let go" and watch them fly!

► Multiple Representations: The Barbie® Bungee Problem

One task that lends itself well to multiple representations is the Barbie Bungee task. There are several versions of this task available, but I have adapted the task as presented on the Illuminations website. In the Barbie Bungee task, students are given materials (a Barbie doll, a meter stick, graph paper, rubber bands) and the basic task: Find out how many rubber bands you would need to create a safe, exciting bungee jump for Barbie from any given height. Students in one class using this task had some experience with t-charts and graphing as well as linear equations and functions, but were not given specific instructions regarding the representations they could use. Most groups set to work connecting rubber bands and sending Barbie leaping off desks, but many were confused about the generalized task: how to find the number of rubber bands required for *any given height*.

The students were given some time to struggle with the problem, and then the teacher approached a group that appeared to be stuck.

Teacher: So, what have you discovered?

Student: We found out that 4 rubber bands is best off of the desk.

Teacher: What have you found out about any given height?

Student: I don't get it. What does that mean?

Teacher: Remember, tomorrow we're going to have Barbie jump off a high place. How can we figure out how many rubber bands to use without knowing the exact height right now? Can you think of a way we used to make a prediction or find a pattern so we can quickly figure the number of rubber bands once you're given the height tomorrow?

Student: We made a lot of t-charts to look for patterns. Can we do that?

Teacher: Try it. First think about what your input and output might be. Then do some more trials with Barbie.

This group got to work making a table of values and measured Barbie's fall for different numbers of rubber bands.

Another group had started by measuring the stretch of each rubber band and figured that each (medium) rubber band stretched to approximately $2\frac{1}{2}$ inches.

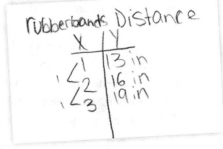

Teacher: Show me what you've found so far.

Student: When you stretch the rubber bands connected together, they come out to about $2\frac{1}{2}$ each.

Teacher: So how would you figure the number of rubber bands needed for a jump from any height?

Student: So, like, for 14 rubber bands, she would fall (pause to calculate) 35 inches.

Teacher: Can you make a sketch to explain how that would work?

In this case, as soon as the students started sketching the problem, they realized they had failed to take Barbie's height into account. When the teacher returned, the group had modified its model.

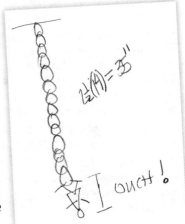

Teacher: What did you discover about your model when you drew the picture?

Student (laughing): We realized Barbie would've cracked her head open! We forgot to add on Barbie's height.

Teacher: That makes more sense. Now, you might want to test your prediction. Try it out from a couple of heights, and then see if you can explain what you just told me in writing.

The teacher was able to validate the correction of the pictorial model and encourage the students to test their reasoning before putting their explanation in writing—another representation.

By analyzing students' initial reasoning and questioning them about their thinking, this teacher was able to direct them to a representation that might be appropriate. Even when students had demonstrated their reasoning using one representation, they were encouraged to test it or represent it using at least one other. Through this process, students were learning how better to organize their thoughts, articulate and defend their reasoning, identify any flaws in their thinking, and adjust their models and solutions where needed.

Formulating the Model and Collecting Data: The 1,000 Paper Cranes Problem

Let's return to the 1,000 Paper Cranes problem. In Chapter 3 students identified the unknowns (how long will it take to make 1,000 origami cranes, how long will the string of 1,000 paper cranes be, and where will we hang

The Mathematics of Origami

Origami reflects many mathematical concepts including symmetry and deductive reasoning, geometric constructions (e.g., two points on a line, bisecting an angle, perpendicular lines), advanced algebra concepts (e.g., construction of conic sections, quadratic equations, squares and cubes of numbers), and trigonometry. Origami can encourage mathematical thinking and precision.

There are many resources for learning and teaching origami. I have found it helpful when I teach my students how to fold a paper crane to have an online demonstration projected at the same time (I like www.wikihow.com /Fold-a-Paper-Crane). I use one class period to teach the students to fold the cranes, discussing the mathematical concepts involved as we go (identifying symmetry; reviewing definitions of squares, rectangles, kites, triangles; defining perpendicular, parallel, bisector). As we fold, I ask students to give me one true mathematical statement about the resulting figure and to justify their statements. They may respond with statements such as, "This is a rectangle. It has two sets of parallel sides and right angles," or "This is a kite. It has two pairs of adjacent sides that are equal in length," or, "This fold bisects this one because it cuts it in half." Students have ample opportunity to practice, and I send them home with origami paper (thinner than regular paper, colored on one side and white on the other) to practice at home. When students return the next class period, they are ready to progress with the task.

the cranes), and identified some helpful problem-solving strategies (break the large problem into a smaller problem, access prior knowledge and similar problems), and decided what information and data need to be collected. At this stage in the modeling process, students formulate the model and collect data in order to take steps toward solving the problem. Many approaches and many solutions to this problem will present themselves. Students have autonomy in making decisions regarding their strings of cranes.

At this point, students will need to make origami cranes in their groups. As acknowledged in Chapter 2, some tasks are more time-consuming than others, and this one will take some time! Still, I have found the task to be incredibly rewarding and engaging for students. In many school settings, the cross-curricular nature of this task may make it possible to involve the art class in teaching the origami process and producing the cranes. But even if that is not the case, the art and act of folding origami in and of itself is highly mathematical, rich, and well worth the time investment in a math classroom.

On Day 1, after making predictions and identifying the unknowns, students begin making cranes, I give students origami paper, walk them through the steps of constructing the cranes (see sidebar), and give them ample time and paper to practice. When students return to class on Day 2, they begin formulating their models and collecting data. Again, depending on their experience with multiple representations, they may need more or less direction at this stage. While on earlier tasks the teacher

Modeling with Mathematics

provides direct instruction in using manipulatives, drawing the scenario, making tables, drawing graphs, or making equations, as the students gain experience on these tasks they will require less and less specific instruction.

Once students have become fairly proficient in making cranes, the next step involves determining the amount of time needed to produce their cranes by timing each other using stopwatches or cell phones. In this portion of the task, students usually display a measure of confidence and autonomy and require little direction.

Students' approaches to determining the time needed to make 1,000 paper cranes vary. In one class, some began by timing each group member separately, then quickly determined they were wasting time. Others folded a few cranes first, reasoning that they would produce cranes faster with practice. Initially, some students found their group's total time (adding the individual times together), and divided this number into 1,000. In this case, the teacher redirected the students with some questioning techniques: "Is this accurate? Would you really make the cranes one person at a time?" Some students decided to use the longest individual time, and others used the students' average time. They then used some sort of proportional thinking to determine the time needed for their group to make 1,000 cranes.

GROUP A

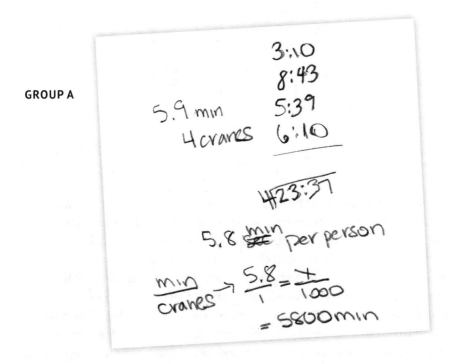

Group A found the average time for their four group members to make one crane (approximately 5.8 minutes per person), and set up a proportion finding the number of minutes for 1,000 paper cranes. They did not, however, consider the fact that they would have four members working at once. This error became evident when we compared results of the groups and performed the next step of modeling—comparing the solution with reality.

GROUP B

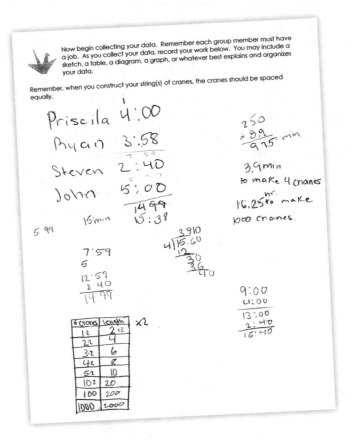

In Group B, students used proportional reasoning differently. In this case, they also determined the average time for their group members to make one crane (3.9 minutes), but then calculated how long it would take each person to make their portion of the total—250 cranes. This group also converted units from minutes to hours in order to begin the next step—comparing with reality.

On the timing portion of the task, even students with limited experience in modeling tasks will typically display some autonomy, understanding that they need to determine the time it takes to produce a crane (or for a group to produce 4 cranes), and use some sort of proportional reasoning to determine the time needed to make 1,000 cranes.

After determining the time needed for their groups to produce 1,000 cranes, students began stringing their cranes in order to determine the length of 1,000 cranes. They measured the string of 5 or 6 or 10 cranes and began recording data. Some students started with a sketch of their work, and were encouraged to look for patterns in their sketches in order to determine the length of 1,000 cranes. Others recorded their measurements in tables they devised. Some groups assumed that this task, like the timing portion of the task, merely involved a direct proportion, which required some redirection.

The group above constructed a string of 5 cranes and used proportional reasoning to determine the length of 10 cranes (by multiplying by 2), and then 1,000 cranes (by multiplying the result by 100). The teacher asked this group to check their work by stringing and measuring 10 cranes, and then to discuss what further information they might need to consider when stringing 1,000 cranes. The teacher also suggested they look for patterns that would help them determine the length of the string with *any number of cranes.*

The students were able to generalize their work, as long as "any number of cranes" was a multiple of 5. Later on, in whole-class discussion, the class further refined the group's thinking to determine the length of any number of cranes (1.8 inches per crane, plus 12 inches of string).

Another student first sketched the arrangement of cranes that her group had decided on in the prior stage—cranes arranged vertically, with 2 inches of string between each crane. They determined that 4 cranes would be 14 inches long, then created a string of cranes to match the picture. The group then discussed where to go from there:

Student 1: If 4 cranes are 14 inches, we could just multiply by, wait a minute, 1,000 divided by 4 is 250, so we could do 14 times 250.

Student 2: So, all together, we would have 250 times 14 inches. That's 3,500 inches for all the cranes.

Student 1: But we don't have just 1 string. We want to make, like, 50 strings.

Student 3: Then that would be 50 strings of 20 cranes, so ... (calculating) that's 14 times 5—70 inches for each string. Plus we have to add some to the top of each one to tie it up. Maybe a foot? So we add 50 feet to the 3,500 inches to get the total!

The group used proportional reasoning to determine the length of the string of cranes. Again, the group was then asked to find a pattern that would help them determine the length of a string of *any number of cranes.*

Rather than sketching first and constructing later, most students started constructing their strings of cranes *before* recording their measurements. When they had a string of several cranes constructed, they began measuring and recording. The students in this class had recent experience using tables, and finding patterns in tables, so many groups organized their data that way.

As noted earlier, many students will view the length of the 1,000 Cranes task as a direct proportion, but depending on their arrangement of cranes, it may not be. Look at the student work to the right:

This student determined that her group's cranes were spaced evenly and that the length of the string increased by 3 inches per crane. She then used this information to determine the length of a string of 1,000 cranes ("Multiply the number of cranes by 3"). This student also converted units (from inches to feet) in order to compare with reality.

# of cranes	Length
1	3 in
2	6 in
3	9 in
4	12 in
5	15 in
10	30
1000	3000

$$12 \overline{)3000}$$
$$250 \, ft$$

The string is a little bit smaller than a football field

The group whose work is shown here, on the other hand, found that their measurements did not result in a direct proportion. The group discussed the pattern on the table:

Student 1: The length goes up by 4 every time, so we can just add 4 over and over.

Student 2: We're not going to do that 1,000 times!

After some guess-and-check, as well as sketching their work, the group determined that they could multiply the number of cranes by 4 and subtract 1 to find the length of the string (and they also discovered their cranes were not equally spaced, so they adjusted the later measurements). They used their pattern to determine the length of a string of 1,000 cranes.

This group's pattern was a little more complicated, and although they recognized that each crane increased the length by $2\frac{1}{2}$ inches, they had trouble finding a general rule for *any number of cranes*.

# of cranes	Length
1	3 in
2	7 in
3	11 in
4	15 in
5	(19) $18\frac{1}{2}$ in
6	(23) 22 in

Pattern:
1×4−1=3
2×4−1=7
3×4−1=11
4×4−1=15
so
1000×4−1 = 3999 in

Team To do ✓

Tables for cranes

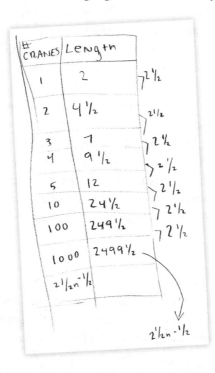

# CRANES	Length	
1	2	⌐ $2\frac{1}{2}$
2	$4\frac{1}{2}$	$2\frac{1}{2}$
3	7	⌐ $2\frac{1}{2}$
4	$9\frac{1}{2}$	$2\frac{1}{2}$
5	12	$2\frac{1}{2}$
10	$24\frac{1}{2}$	⌐ $2\frac{1}{2}$
100	$249\frac{1}{2}$	⌐ $2\frac{1}{2}$
1000	$2499\frac{1}{2}$	
$2\frac{1}{2}n - \frac{1}{2}$		

$2\frac{1}{2}n - \frac{1}{2}$

Student 1: Since they're $2\frac{1}{2}$ apart, we just multiply by $2\frac{1}{2}$.

Student 2: It doesn't work. One times $2\frac{1}{2}$ isn't 2.

The teacher suggested they sketch their work to see if they could find the patterns. The group discovered on their picture that the distance from one crane to the next was $2\frac{1}{2}$ inches, but that the distance to the first crane was only 2 inches:

Student 1: I see where the $2\frac{1}{2}$ comes from, but what do we do with the 2?

Student 2: The string at the top is shorter. That makes the 2.

Student 3: If it's $\frac{1}{2}$ inch too short, we can just take away $\frac{1}{2}$.

Student 1: So times $2\frac{1}{2}$ and minus $\frac{1}{2}$?

The students checked their work on several measurements and found that it worked. They, too, discovered a formula to help them determine the length of a string of 1,000 cranes. In the next stage of modeling, students will have to convert units from inches to feet or yards in order to better compare with reality.

Although some students can find patterns directly from the table, by combining the table and the sketch students can make important connections and move up the ladder of abstraction. By the time students get to a formula (e.g., $L = 2\frac{1}{2}n - \frac{1}{2}$), they may understand that it works with the data but loses the connection with the concrete model (in this case, the cranes and the string). By sketching the work and looking at the physical model, students can step back down the ladder of abstraction and connect the numbers in their formula with the physical model or sketch (the $2\frac{1}{2}$ is the distance from one crane to the next, and the $\frac{1}{2}$ is subtracted because the initial string is $\frac{1}{2}$ inch shorter). Even when students construct a successful model using a particular representation, it is helpful to compare several models to deepen understanding.

Students collected the data and determined the correct models for calculating the time and length, then proceeded to solve the problems in the task But the solution is not the finish line. The next chapter considers the reasonableness of the solutions, and shows how to direct students to examine whether their solutions make sense in context.

Conclusion

The second stage of mathematical modeling involves formulating the model and collecting data in order to find a solution to the given problem. You can encourage your students to try a variety of representations to determine the best method(s) to solve the problem. When your students become accustomed to multiple representations, their increased mathematical autonomy will better equip them to approach a problem from a variety of standpoints. They will discover that the most complicated method for solving a particular problem may not always be the best one, and that their individual learning styles may favor one representation over another on a given problem. When you offer your entire class various methods of representing and solving a given problem, they are likely to see connections between the representations, and, perhaps, understand that another method may have been more efficient or accurate than their own. When they move on to the next stage of modeling—interpreting their solutions and comparing them with reality—they may adapt or adjust their model to better represent the problem and its solution.

Implementing rich modeling tasks and multiple representations presents a challenge to teachers, who must themselves develop a deep understanding of the different methods and the layers of the problem, but it also opens an exciting new way of teaching and learning. When you begin to incorporate these tasks into your classroom—when you give your students the opportunity to try different methods, make decisions about representations and problem solving, and look at problems from different viewpoints—you will be rewarded by the engagement, motivation, and depth of understanding they reveal. Each time I use these tasks, I learn more about my students and their thinking, as well as more about the problem and the mathematics, than I knew before.

Interpreting the Solution and Comparing with Reality

Why, sometimes I've believed as many as six impossible things before breakfast.

—LEWIS CARROLL, ALICE IN WONDERLAND

A friend of mine came to me recently, saying, "I'm going to pull my hair out!" because she was trying to incorporate math into her science class. She had been teaching a lesson comparing resting heart rates to exercise heart rates ("Your heart pumps approximately 1.3 gallons per minute when at rest. How much blood does it pump when you are exercising?"), and her students came up with a variety of ludicrous (her word) answers. Some of her students concluded that the active heart would pump more than 14,000 gallons per minute! Others came up with negative answers. Despite directed questions ("Does this make sense? Could you fill a swimming pool in a couple of minutes with the blood pumped through your body?"), some students stuck by their answers, reasoning, "That's what the calculator said." This friend was less frustrated by the students' errors in calculation than by their complete lack of connection between the math and the real problem.

Modeling with Mathematics

Anyone who has taught word problems or applications problems has run across the same frustration. Our students sometimes get so wrapped up in the arithmetic involved that they suspend reality when solving these problems. This may be understandable when the problems involve concepts with which they are unfamiliar (e.g., pressure on a scuba diver or the vibrations of a violin string), but is particularly irritating when the concepts should make some sense to them ("Mary drives 120 miles in 3 hours. How long would it take her to travel 300 miles, if she travels at the same rate?" or "A $50 jacket is on sale for 20 percent off. Find the sale price").

The Importance of Interpretation

In a traditional math exercise, the calculated solution is the final goal, but in mathematical modeling, there is still more work to be done: interpreting the solution and comparing it with reality. We are often guilty of believing that the highest rung on the ladder of abstraction (the abstract concept, the formula, the equation, or the solution to the calculations) is the ultimate goal. Modeling, however, reminds us that we must, when the calculations are complete and the solution is found, climb back down the ladder to the concrete problem at hand. We must interpret the solution in light of the problem, and we must compare the solution with reality—ensuring that it makes sense in the context of the problem.

When students consider the reasonableness of their solutions, they can begin to make powerful connections between their own problem solving and the real world, and to bridge the gap between classroom mathematics and their real lives outside the classroom.

➤ Unrealistic Word Problems and Unrealistic Answers

Traditional math instruction in the United States has given little consideration to the realistic constraints of word problems, and these problems in the traditional textbook do not typically call for students to consider the reasonableness of their solutions (De Corte et al. 1995; Greer 1997; Inoue 2009). Too often, word problems represent an extension of a procedure students have practiced in numerous exercises, with little attention to sense making in light of the context of the problem. It is important to look at the way students typically view and solve word problems to help them extend their thinking to consider the context of the problem and the reasonableness of their solution.

Let's look at an often-cited problem from the math education literature (Baruk 1985). This problem was given to primary students in order to assess their approach to word problems:

> Twelve sheep and 13 goats are on a boat. How old is the captain?

This may seem like an extreme, if not ridiculous example, but the majority of the primary grade students answered "25 years old"! They gave little attention to the context of the problem, and assumed that all of the facts needed were given. Granted, most of our textbook problems are not this nonsensical, but the Captain and the Goats problem shines a light on the way our students often look at word problems. Through repeated exposure to traditional textbook word problems, students know the problem will probably require them to use the operations or algorithm they have been practicing earlier in that same lesson, and they "solve" the problem using that algorithm with little regard for its context or the reasonableness of the answer.

Many textbook word problems are so removed from the real-life experience of students, or their understanding of the world around them, that students must suspend their sense of reality and put aside their prior knowledge in order to solve these problems. Jo Boaler describes these problems as using "pseudocontexts," where students are required to step into the alternate reality of "Mathland" in order to solve them. Here is an example of a typical textbook word problem from Boaler's (Boaler and Staples 2008, 51) book:

> A pizza is divided into fifths for 5 friends at a party. Three of the friends eat their slices, but then 4 more friends arrive. What fractions should the remaining 2 slices be divided into?

In Boaler's Mathland, students must put aside their understanding of the way the world really works in order to solve the problem. In the example given, for instance, students know that if 4 more friends arrived, they would order more pizza or find something else for them to eat rather than divide 2 slices among 6 friends! When we ask students to solve these pseudocontextual problems, we do not give them the opportunity to be interested and engaged in real problem solving. We do not ask them to make connections between the real world and mathematics, making them less likely to see classroom mathematics as relevant to their lives. We also do not require them to check the reasonableness of their solutions, because their solutions are *not* reasonable (would we really serve our late friends $\frac{1}{3}$ of a slice of pizza?).

Other problems demonstrate the fact that, even when the problems make sense, students pay little attention to the meaning of their solutions. On the Third National Assessment of Education Progress (Carpenter et al. 1983), thousands of thirteen-year-olds were presented with the following Army Bus problem:

Modeling with Mathematics

An army bus holds 36 soldiers. If 1,128 soldiers are being bused to their training site, how many buses are needed?

Only 24 percent of the students tested solved the problem correctly. Seventy percent of the students performed the calculations correctly (finding the answer "31 remainder 12," $31\frac{1}{3}$, or $31.\overline{3}$), but only about one-third of those students understood that they would actually need 32 buses. The majority of the students did not demonstrate understanding of the unreasonableness of their answers (Verschaffel and De Corte 1997). The problem exemplifies the need for students to examine their solutions in terms of the context of the task. The first step in this process is for students to understand what their answers mean.

Interpreting the Solution

When students are solving even routine textbook word problems, it is vital to teach them to determine what their solution means. As calculations get more complicated, and the arithmetic takes students further away from the original problem, students become less likely to remember what they were trying to figure out in the first place. Students need to become accustomed to asking themselves, "What did I just figure out?" and including appropriate units on their solutions. When students become accustomed to including units (e.g., inches, liters, dollars, puppies) as they work through word problems, they will begin to attend to the meaning of the problem rather than merely the calculations.

On the Army Bus problem, for instance, if students had identified and kept track of the units during the problem, and reported units on the answer, they may have been more likely to understand that $31\frac{1}{3}$ buses does not make sense. If students had used a brief key when they set up the problem, they might have been more likely to recognize the unreasonableness of their answer:

$$\frac{\text{buses}}{\text{soldiers}} \longrightarrow \frac{1}{36} = \frac{?}{1128}$$

➤ Explaining and Justifying Your Solution

One way to ensure students understand the meaning of their solutions is to ask them to express their answers to word problems, problem-solving opportunities, and modeling tasks in complete sentences (or paragraphs when appropriate). Consider the following Tree Shadow word problem given to a group of sixth-grade students:

At the same time of day, you cast a shadow that is 8 feet long, and a tree casts a shadow that is 32 feet long. If you are 5 feet 3 inches tall, how tall is the tree?

Some students converted all units to inches before calculating, and others converted to feet. Students who correctly completed the calculations had answers that included "21," "252," "252 feet," "21 inches," "21 feet," and "252 inches." Even on this relatively straightforward question, it is unclear whether the students remembered what they were calculating and whether their solutions made sense to them. When students become accustomed to expressing their answers in complete sentences, they are more likely to convey their understanding (or lack thereof), and engage in some sense making regarding their response. On the Tree Shadow problem, this could simply mean expressing the solution as, "The tree is approximately 21 feet tall." Although students may not be able to visualize 252 inches, they are more likely to understand the concept of 21 feet.

In a discussion of the problem, ask students whether their answer makes sense. The students in this classroom responded, "Yeah, a tree can be 21 feet tall," "I am shorter than my shadow, so it makes sense that the tree is shorter than 32 feet," and "The tree's shadow is 4 times my shadow, so it makes sense that the tree is 4 times my height. That would be a little more than 20 feet." Taking the time to have a brief discussion about "Does this make sense?" helps students make important connections between the classroom and the world around them, and increases their engagement in the problem solving. In addition, it helps students with incorrect answers understand why their answers do not make sense.

Using Incorrect Answers

In the Tree Shadow problem, several students set up the proportion incorrectly and got a solution of (some variation of) approximately 48.76 feet. Ask students who get the wrong answer if they are willing to share how they know their answer was incorrect. When a classroom has truly built a culture of inquiry, and students understand that we learn from our mistakes, some students will be willing to share. In this case, one student remarked, "Now I see it. The tree should be shorter than its shadow, so 48 feet doesn't make sense." As the students explain what their answers mean and reflect on their solutions—whether correct or incorrect—they engage in sense making and use deeper reasoning skills rather than mere rote calculations.

Whether or not your students are willing to share their mistakes, it is always a powerful learning experience to ask students to find the error on a problem or compare the work of two imaginary students. You can use common errors you have seen on student work or assessments to create an opportunity for students to study the work of others, find and explain their errors, and correct them.

The following Mixing Paint problem was used with a seventh-grade class:

> Luz is mixing paint. She wants to make lime green paint, but she only has primary colors (red, blue, and yellow). She knows that she can make lime green by mixing equal quantities of yellow and green, and that she can make green by mixing equal quantities of blue and yellow. If she wants to make 12 cups of lime green paint, how much blue paint and how much yellow paint will she need?
>
> Look at Luz's work below. Did she calculate correctly? Does her answer make sense? How do you know? If she is incorrect, correct her errors.

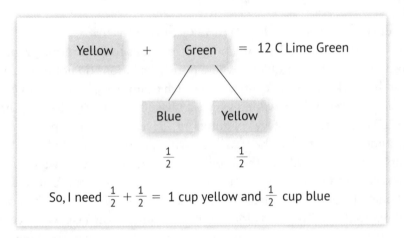

This is a difficult problem for seventh-grade students to do on their own (fewer than 50 percent of nearly 12,000 students tested answered correctly on a very similar task) (Noyce Foundation 2012), but even students for whom the task may be difficult can analyze the work of another student and assess the reasonableness of the solution.

Students in one seventh-grade class immediately noticed that Luz did not have enough paint to make 12 cups, that she had too much yellow paint, and that the total fraction for the whole was $1\frac{1}{2}$ rather than 1. When students discussed and understood some of Luz's errors, they were motivated to solve the problem themselves, avoiding some of her mistakes. Some students approached the problem using a different approach than Luz, and some who did not know where to begin used Luz's original approach with corrections.

By examining the work of others, especially through the process of finding others' errors in thinking, students begin to examine the reasonableness of their own solutions rather than simply giving an answer and moving on. Another strategy that is useful for encouraging students to examine the reasonableness of their solutions is the use of nonroutine and counterintuitive problems.

➤ Nonroutine Problems

The math education research presents many interesting examples of students' unreasonable solutions to nonroutine problems. I have found that sharing some of those with my students before they engage in their own modeling of nonroutine tasks encourages them to seek to avoid the same sort of mistake. Guiding students to assess the reasonableness of solutions in nonroutine problems is a productive exercise, and you can apply any number to strategies to investigate these problems with your students.

When Pure Calculation Doesn't Make Sense

One way you can help students understand solutions to nonroutine problems is to discuss a variety of problems where pure calculations do not produce a solution that makes sense in context. One such example is, "If one man can dig a hole in 60 seconds, how long would it take 60 men to dig the same hole?" (Inoue 2009). In the early stages of autonomy development, just pose the question to students to see what they come up with. Some students will certainly come up with "1 second," but there will usually be a student or two who will express skepticism about this solution with comments such as, "You wouldn't have 60 men digging a little hole," or "All those shovels in the same hole won't make the work faster." A discussion about the reality of the task will encourage students to begin to consider the context of the problems they are solving.

Greer (1993) cites research with some more complicated problems. For instance, he posed the following problem to students: "An athlete's best time to run a mile is 4 minutes 7 seconds. About how long would it take him to run 3 miles?" In his study, most students responded with a solution assuming that a direct proportion would be an appropriate estimate for the athlete (although, interestingly enough, students using calculators tended to make errors confusing the decimal representation of the 7 seconds). Very few students expressed the fact that it would be extremely unlikely for the athlete to keep up the same pace for 3 miles. It may be helpful to pose this problem to students, and contrast the reasonableness of the solution with that of a similar problem: "A barge travels a mile in 4 minutes and 7 seconds. About how long would it take to travel 3 miles?" Exposing students to similar problems in which the direct proportion is more appropriate (the barge) and less appropriate (the athlete) highlights the importance of considering reasonableness and the context of a solution.

Listing Assumptions

Another method for helping students interpret the solution to a modeling task and compare it with reality is to ask them to list some assumptions they are making when providing a solution that is a result of pure calculations. For instance, on a variation of the problem cited earlier ("An athlete's best time to run a mile is 4 minutes 7 seconds. About how long would it take him to run 3 miles?"), Inoue (2009) suggests that students might think the assumption is that the runner can keep up the same pace for the entire race because he is a trained athlete (very unlikely if students follow track and field), or that the total distance is not run continuously, but 1 mile at a time.

Another example involving rate-time-distance is a variation on an Inoue problem (2009):

> You need to be at the airport at 7 p.m. You leave home at 5 p.m., and the airport is 120 miles away. You travel the first 60 miles in one hour. Will you make it to the airport on time?

If we ask students to list some assumptions along with their calculated solution, we can encourage them to build that bridge between the math classroom and their daily lives. In this case, they might say something like, "I can make it to the airport on time if I travel at the same rate. That assumes I do not run into an accident or a lot of traffic."

Encountering Problems with Counterintuitive Answers

If we can occasionally provide students with problems that have counterintuitive solutions, we spark their interest and curiosity and require them to think about their solutions. Look at the BMX Bike Race problem, a task I developed based on the interest of some of my students:

> Mario and Antonio are in a BMX bicycle race. The race is twice around the $\frac{1}{2}$-mile long BMX track. In Mario's first lap around the track, he averages 16 mph, but in the second lap, he has some problems with his bike and averages only 12 mph. Antonio averages 14 mph on both laps of the race. Who won the race?

The majority of students will assume that the race ended in a tie (because they assume Mario's average rate is also 14 mph). At this point, we ask the students to justify their assumption mathematically.

After working on the calculations, students realize that Mario takes 4.375 minutes to complete the race (or 4 minutes, 22.5 seconds), and Antonio will finish the race in approximately 4.286 minutes (or approximately 4 minutes, 17.14 seconds). So, contrary to the students' assumption, Antonio wins the race!

One possible calculation:

MARIO	ANTONIO
Lap #1: $\dfrac{\text{miles}}{\text{minutes}} \to = \dfrac{16}{60} = \dfrac{0.5}{n}$ $n \approx 1.875$	$\dfrac{\text{miles}}{\text{minutes}} \to \dfrac{14}{60} = \dfrac{1}{n}$ $n \approx 4.286$
Lap #2: $\dfrac{\text{miles}}{\text{minutes}} \to = \dfrac{12}{60} = \dfrac{0.5}{n}$ $n = 2.5$	Total time: 4.286 minutes
Total time: 4.375 minutes	

This solution is counterintuitive, and a great deal of the thinking is done *after* the calculations. Students will express surprise at this unexpected answer, and see the power and excitement of the problem as students figure out and explain *why* the answer is not what they expected.

In one classroom, the explanations included the following:

"Mario really only went about 13.7 miles per hour if he went one mile in 4.375 minutes. Antonio went 14 miles per hour the whole time."

"When we go for the same amount of time, both boys go the same distance, but when we go the same distance, the times are different. Like, in 2 hours, both boys would go 28 miles, but for, like, 48 miles, Mario would take 1.5 hours for the first part (24 miles), and 2 hours for the second part, so 3.5 hours total. Antonio would take 3.43 hours, so he goes faster."

Although these calculations were correct, the teacher did not believe that her students clearly understood *why* Antonio won the race. She prompted further discussion:

Teacher: I still don't understand why Antonio won. Can you keep working on this so you can explain it to me?

Student: What do we do now?

Teacher: You said that when the boys go the same distance, their times are not the same. Can you use "friendly" numbers, or a picture or table, so we can understand it better?

Student: What if they go, like 100 miles?

Teacher: Try that.

The students kept working. One group reported their findings:

> Okay. Mario takes 6.25 hours to go the first hundred miles, and 8.33 hours to go the second hundred miles, so 14.58 hours to go 200 miles. Antonio takes 7.14 hours to go the first hundred miles and 7.14 hours to go the second hundred miles, so 14.28 hours to go 200 miles. Even though Mario is ahead at 100 miles, it takes him a lot longer for the last 100 miles, so Antonio passes him and wins.

Another group made a table comparing the times it took each boy to reach a certain point in the race (they, too, made a 200-mile race with 100 miles on each lap). The table looked like this:

MILES	MARIO'S TIME (hours)	ANTONIO'S TIME (hours)
100	6.25	7.14
120	7.92	8.57
140	9.58	10.00
160	11.25	11.42
180	12.91	12.86
200	14.58	14.29

They explained, "Even though Mario was ahead after the first lap, he started going a lot slower. Antonio was going faster, and passed him at about 180 miles. That's why he won."

A third group, nodding, reported, "Mario went fast on the first lap, so he spent more time going slowly—at 12 miles per hour—than Antonio. Antonio went 14 miles per hour the whole time."

When students face a problem that has a counterintuitive solution, discussing and investigating the solution can expand and enrich their mathematical experience.

Another problem having a counterintuitive solution came out of a real experience that I shared with my students. In California, as in many other parts of the country, the past several years in education have been fiscally challenging. In my district, teachers had not gotten a pay raise for 9 years. That year, however, many districts finally got raises. I posed the following scenario to my students:

> Our district just received a 2 percent (retroactive) pay raise for this year, and approved a 3.5 percent pay raise for next year. My daughter, who is a teacher in a nearby district, received a 2.5 percent (retroactive) pay raise for this year, and a 3 percent pay raise for next year. Which district got the better deal?

Once again, the obvious answer is that all teachers got the same deal. Nearly every student came up with some variation of, "It doesn't matter. They'll all end up with a 5.5 percent raise." I asked them to justify their answers mathematically. Many were initially stumped as to how to begin until a couple of students started asking, "Does it matter what your starting salary is?" I suggested they make up a starting salary to answer their own questions. This prompted them to begin with a starting salary (and it was fascinating to find out the students' varying ideas about teacher pay), and to find a 5.5 percent increase of that initial amount. Then they broke into small groups and began investigating the two districts' options. Once again, as the students began exploring the problem, it was gratifying to hear their thoughts and see their surprise as they began to understand why their initial predictions were incorrect. The discussion points started as follows:

> "It's not the same as getting 5.5 percent when you split it up into 2 years. You get more this way."

> "We did $10,000. If you just get a 5.5 percent raise, you'd make $10,550. In our district, you'd make $10,557, and your daughter would make $10,557.50." (I'm not sure whether this group thought I made $10,000 a year or $10,000 a month!)

> "We started with $50,000 and got $52,750 the first way, $52,785 for you, and $52,787.50 for your daughter. Her district is better."

Some students volunteered to share their calculations so that the class could see how they reached their conclusions. Then I asked them *why* it makes a difference if the percent raise is split in two, and how it is split. Again, this sparked a lively discussion. Some students immediately understood the basic concept ("you're getting a raise on top of your raise"—a good introduction to the concept of compound interest), while others saw the differences but did not understand why. Some groups decided to investigate what would happen if the 5.5 percent were split differently (and more dramatically). One group shared the following:

> We did 0.5 percent the first year, and 5 percent the second year with $60,000, and then we did our district—2 percent first and then 3.5 percent. We got $63,315 the first way, and $63,342 the second way. If you get the raise all together, you just get $63,300, but if you get the smaller raise first, you're getting the other raise on a higher salary.

Many students seemed confused, so I asked this group to show us their work on the board. Their table clarified their explanation for the entire class:

Starting salary is $60,000

SALARY	PLAN A: 5.5% ONE TIME	PLAN B: 0.5% THEN 5%	OUR DISTRICT 2% THEN 3.5%	DAUGHTER 2.5% THEN 3%
This year	$63,300	$60,300	$61,200	$61,500
Next year	$63,300	$63,315	$63,342	$63,345

They explained that teachers in districts that got a percentage increase on the larger salary (after year 1) ended up with the larger pay raise. As an extension of this activity, I encouraged students to continue the investigation as a Problem of the Week, to see if we could extend our understanding of this problem and make any new discoveries. The complete evaluation of the optimal pay raise arrangement is actually a fairly complicated quadratic function (see the sidebar for more information), and continued investigation will reveal to students that, at some point (2.75 percent the first year and 2.75 percent the second year), the salary increase reaches a maximum, and begins to decrease again.

When we give students nonroutine and counterintuitive problems, we task them with thinking and reasoning critically. They are motivated to "prove" they are correct, or that another student is incorrect. If we stop every problem at the conclusion of the calculations, students miss out on these powerful learning experiences and are far less likely to connect their lives outside of the classroom with the work they are doing in their math classes.

Optional Teacher Raises—Extended

x = salary increase (in decimal form) Year 1

$S(x)$ = Salary Year 2
For $60,000 at a total of 5.5%

$S(x) = \$60{,}000(1 + x)\,[1 + (0.055 - x)]$
$S(x) = \$60{,}000(1 + x)(1.055 - x)$
$S(x) = \$63{,}300 + 3{,}300x) - \$60{,}000x^2$

Maximum: $(.0275, \$63{,}345.375)$

So, the maximum salary would be achieved if the salary were split into 2.75% the first year and 2.75% the second. It would begin to decrease as the first year's percentage increases beyond 2.75%.

Examples of Students Interpreting Solutions

Let's look back at some student solutions to problems we investigated earlier.

➤ Kicker Ramp

The Kicker Ramp problem from Chapter 2 (page 30) asked students:

> Your brother wants to build 3 different kicker ramps for practicing tricks on his skateboard. The first ramp is 6 feet long, 4 feet wide, and 1.5 feet high. Find the steepness (slope) of this ramp, and then sketch the plans for two other ramps using boards (a) 8 feet long and 4 feet wide, and (b) 12 feet long and 6 feet wide. Decide on the height of the ramp for each, and determine the steepness (slope) of all three ramps. How high do you think you could build each ramp in order to make a reasonably steep kicker ramp? Explain your reasoning.

When I present this problem, I give the students manipulatives (a ruler, some blocks, some pipe cleaners, and quarter-inch graph paper), and they work in groups. Even students who have not had a formal introduction to the mathematical definition of *slope* can understand, visualize, and explain the steepness or slope of a visual object like a kicker ramp. Too often, students learn "rise over run" in pre-algebra or algebra class with no connection to real life. With the kicker ramp, students begin to make informal definitions of slope by finding ways to describe the relative steepness of the ramps. In a sixth-grade classroom, some students' initial discoveries included:

> "We don't think the width of the ramp effects [sic] the steepness of the ramp."

> "The ramp that is 6 feet long and 1.5 feet high is the same steepness as the ramp that is 12 feet long and 3 feet high."

> "For the same length of board, the higher the ramp, the steeper the slope."

As students progress with the task, some will begin to investigate ways to define the steepness of the ramp mathematically:

> "We can measure this angle at the bottom of the ramp to find out how steep it is."

> "We know that 6 feet long and 1.5 feet high is the same as 12 feet long and 3 feet high. So that would also be the same steepness as 24 feet long and 6 feet high or 36 feet long and 9 feet high."

When I asked the students in this class how they could prove that the ramps would be the same steepness, they showed me the "ramps" they had constructed with rulers and blocks. I asked them to show me using math. Several students came up with the idea of equivalent fractions:

$$\frac{6}{1.5} = \frac{12}{3} = \frac{24}{6} = \frac{36}{12}$$

The students used logic, physical models, and mathematics to justify their answers.

In grade 7, students typically study a more formal definition of slope. For the Kicker Ramp problem, students are not given the measurement of the "run," but can measure the manipulative model (with 1 inch being a scale model for 1 foot), or the scale on the graph paper to estimate the length of the "run." These students can provide a more accurate measure of the slope of the ramps (using either a fraction, a decimal, or a percentage). Grade 7 responses included:

> "The first ramp has a slope of about 26 percent because the rise is 1.5 and the run is about 5.8, so 1.5 divided by 5.8 is about .26."

> "We made the second ramp 8 feet long and 4 feet high so the bottom side was about 7 feet. That makes a slope of $\frac{4}{7}$ or 57 percent."

> "If the board is 12 feet long, we can make it 1 foot high. That makes a flatter ramp with a slope of about 8 percent."

Tasks like the Kicker Ramp problem give students experience with formal and informal investigation of slope, congruency, and the Pythagorean Theorem. This investigation can be extended to precise and formal definitions in grade 8 and beyond as students work with the Pythagorean Theorem and slopes and can be extended further to make connections with trigonometry.

After students design and sketch the ramps, and determine the steepness (or slope) of each, they must still discuss the solutions in context. As we solved the problem, we climbed the ladder of abstraction, moving further and further away from the actual skateboard ramps (from ramps, to models of ramps, to pictures of the models, to measuring the pictures or the lines, to, perhaps, plugging numbers into a formula to find the lengths of sides). We must now encourage students to step back down the rungs of the ladder to assess the reasonableness of their answers. In this case, they must consider and discuss real kicker ramps. Some students will probably decide that a ramp with a 57 percent slope (or some other, higher slope) might be too steep for skateboard tricks, or that a ramp with a slope of 8 percent (or some other, lower slope) might not be steep enough to gain speed for a skateboard trick. This move back down the ladder of abstraction reminds students of the problem they were solving in the first place and of the questions they need to ask, "What am I trying to figure out in this problem?" and "Does my answer make sense?"

➤ The 1,000 Paper Cranes Problem

At this point on the 1,000 Paper Cranes modeling task, students have solved the basic problems (How long is your string of cranes? How long will it take you to make a string of 1,000 paper cranes?) and must now consider the reasonableness of their solution and use their reasoning to complete the task by making a recommendation to the principal regarding where and how to display the string(s) of cranes.

On the 1,000 Paper Cranes problem, when constructing a proposal for the principal, students had to consider a reasonable way to predict the actual time needed to make the 1,000 paper cranes as well as a reasonable way to display the string(s) of cranes. One group calculated that it would take the group of four 16.25 hours to construct the cranes. They first had to consider whether the answer made sense, and then figure out how that time would be reasonably split. Their discussion went like this:

Priscila: So it would take us 16 hours and 15 minutes to make 1,000 cranes.

John: So we'd each have to work a little more than 4 hours?

Ryan: I don't think that's what we figured. I think we *all* need to work 16 hours.

Priscila: Okay. It takes me about 4 minutes per crane, so I could make (pause for calculation), okay, I could make 15 cranes in an hour. That means I could make 240 cranes in 16 hours. We're making 250 each, so that makes sense. We each have to work 16 hours and 15 minutes.

Steven: But we can't make cranes for 16 hours straight. I think we should say it would take us 2 weeks to make the cranes. That would be a little more than an hour a day. I think we could do that.

These students engaged in sense making by first checking their solution and making sense of their calculations, and then by putting their solution (16.25 hours) into a real-life, reasonable context. It would have been easy to get lost in the crane construction, timing, and calculations, but the group was brought back down to the real-life problem by being asked to present their proposal to the principal in terms that would make sense. (I must add here that, as the discussion of crane making continued, the students reached a consensus that very few of them would actually spend 16 hours folding cranes! They decided as a class that they would include the time required for the whole class to make the 1,000 cranes, under the direction of the winning group.).

Another group considered the reasonableness of their solution and how best to implement it—that is, how to display the cranes. This group had determined that their string of cranes (strung horizontally) would be 3,999 inches. Their discussion went like this:

Ana: So where are we going to display the cranes?

Luis: I think we should put them across the front of the school!

Jacob: I think they should go in the cafeteria. They need to be inside so they don't get wrecked.

Morgan: We first have to figure out how long the string is. How long is 3,999 inches?

Luis: Divide by 12 (calculates). It's 333 feet. Let's look at the cafeteria diagram (Students had had the opportunity to measure various areas of the school). It's 70 feet by 30 feet. If we go all the way around, it would be 70 + 70 + 30 + 30. That's only 200 feet.

Morgan: We could go back and forth across the cafeteria. If we go diagonally, we could go across the middle, like 40 feet each, so back and forth about 8 times.

Ana: That would look pretty. Let's do it that way!

This group converted units to make their solution more understandable and discussed where the cranes would look the best and last the longest. They then considered perimeter measurements before finally deciding on and sketching a plan for displaying the cranes. Other groups had arranged their cranes vertically on the string, and had to consider how to split the string so that the cranes would not hang too low, how many strings they would create, and whether splitting the strings would change the measurements.

In all cases, students were required to put their solutions into context, check their solutions for reasonableness, and make decisions based on real-life considerations. Students who had made earlier errors in calculations realized their errors as they began putting their solutions into context and were able to discover what they had done wrong. Even though each group had come up with its own solution and plan, students compared their solutions with others, and often realized their answers did not make sense. One group, for instance, had calculated their time to make the 1,000 cranes as about 24 minutes per crane (they had added together the individual times of each group member), and so, 400 hours to complete the task. They were dismayed to think that, if they worked for an hour a day, it would take them more than a year! After discussing their solution with other groups, they reconsidered their work, found their error, and corrected it.

Conclusion

One of the most powerful aspects of mathematical modeling reveals itself through interpreting the solution and comparing it with reality. Modeling tasks involve students in making decisions, using multiple representations, and employing a variety of calculations. Often these decisions, representations, and calculations take the students away from the original problem posed, and they sometimes lose the connection to the original problem or fail to consider what the solution means and whether it is reasonable. When we engage our students in the task of interpreting the solution and comparing it with reality, we increase their depth of understanding, enable them to discover their own errors in calculation or reasoning, and help them make important connections between the numbers in the classroom and the world around them.

Communicating and Implementing the Solution

6

Doing mathematics should always mean
finding patterns and crafting beautiful
and meaningful explanations.

—PAUL LOCKHART, A MATHEMATICIAN'S LAMENT

In many traditional math classes, the answer is the end of the story. Once a student calculates the correct (or incorrect) answer, he or she is finished with the problem and moves on to the next. But if students are to truly understand the mathematics they are learning, they must do more than follow rules and procedures to arrive at an answer: they must be able to explain what their answer means. They can only construct meaning of the mathematics they are doing by, "exploring, justifying, representing, discussing, using, describing, investigating, predicting, in short by being active in the world" (Countryman 1992, 2). When students complete the solution stage of a modeling task, it is important that they then communicate their solution in a meaningful way.

Good communication is vital to a successful classroom. Good teachers understand the importance of effective communication when it comes to getting information and content across to students, but we as teachers must also

---

consider the importance of the students' ability to communicate with us and with their peers. We can and must teach students how to communicate their ideas, reflect on their strategies, and justify their reasoning in the math classroom so that they may better share their thinking and clarify understanding for themselves and others. The National Council of Teachers of Mathematics (NCTM) emphasizes the importance of communication in its *Principles and Standards for School Mathematics* (2000, 60), stating, "Students who have opportunities, encouragement, and support for speaking, writing, reading, and listening in mathematics classes reap dual benefits: they communicate to learn mathematics, and they learn to communicate mathematically."

Most teachers would concur that they did not have a true understanding of many mathematical concepts until they had to teach them. Only when they had to explain those concepts to others (their students) did they have to break down the concepts and present them in a manner that was clear and understandable. When we teach students how to communicate their thinking and their solutions, we are not only informing others, but also increasing and reinforcing learning for the communicators as well.

Communicating the Solution

Communicating the solution in the math classroom can take many forms. On a brief mathematical modeling task, communicating the solution might mean merely writing a sentence defining the solution rather than just reporting the numerical answer, or recording units and brief descriptions of work as students proceed. The communication phase might also involve drawing a picture or sketch, making a table, or writing an explanation. In any case, the communication should convey the solution to the problem, reflect some reasoning about the answer, and put the solution into the context of the original problem. When students communicate the solution clearly—that is, in a way that clarifies their thinking and reasoning—they are increasing understanding for themselves and their audience. Even though communication is part of the final stage, it often serves as an additional check for reasonableness along the way and may help other students better understand their own work as they view a problem in a different way.

I recently attended a powerful presentation by Cathy Seeley, the former president of NCTM. Her talk was titled "Preparing Problem Solvers to Think Mathematically," and she presented a short video of a student, Marisa, solving the following problem: "There are 295 students in the school. School buses hold 25 students. How many school buses are needed to fit all of the students?" (http://frizzleblog.scholastic.com/post/be-wary-using-key-words-your-math

-instruction). Marisa solved the problem by adding $295 + 25$, getting a sum of 320. The video is emphasizing the student's reliance on the cue word *all* in the problem, leading Marisa to determine that *all* means addition. Although Marisa pauses for a moment when she reads her solution of 320, appearing to question her answer, she is relying more on the word cue than on her reasoning. If you watch the video, you will see the moment Marisa questions her reasoning. You can almost see the question in her eyes, as if she is asking herself, "Does this make sense?" and then quickly resolves her question, assuring herself that "My teacher told me that *all* means add."

Let's take the example a little further in light of communication. In the video the teacher/researcher asks Marisa, "How did you figure that out?" and Marisa explains, "Since I heard *all*, I figured, oh, that's 'plussing.'" If this had been a real classroom scenario, the teacher might have helped Marisa by merely asking her another question about her solution. She might have asked, "Does this make sense?" or "Can you show me how this works?" If Marisa had been asked to really communicate her solution in a complete sentence, and, perhaps, justify or illustrate her solution, she would have been more likely to see her error and correct her misunderstanding. She would understand that it is not reasonable to have 320 buses for 295 students, and she may have even extended her thinking to understand that *all* does not always mean addition.

If Marisa had been given this problem on a traditional assessment, we might have marked it wrong and moved on, but when we focus on communication, we can turn an error into an opportunity. Merely putting the solution back into context, using very few words, might be the little nudge a student needs to recognize her mistake.

Marisa's story helps us understand the importance of communication in our classrooms. Some students (and some parents) might express frustration when we insist on, for instance, writing the solution to a word problem, problem-solving task, or modeling task in complete sentences. I think the simple example of Marisa and the buses illustrates the importance of communicating the solution as well as the teacher's role in communicating back to the student when considering the answer. This does not have to be an arduous task. It may require students to write a few words rather than just a number, or require the teacher to ask one or two additional questions. As students learn to properly communicate their solutions to a problem, we might just walk around the classroom asking a few guided questions: "Does your answer make sense?" "How do you know that is true?" "Can you show me why that works?" or "Can you show me in another way?" Whether the students' solutions are correct or incorrect, this kind of reflection and communication can deepen understanding and help students recognize their errors.

On more complicated modeling tasks, we might need to remind students to use written or numerical notation or units on each part of the problem as they proceed. In some cases, students may get so wrapped up in the calculations that they forget what they were figuring out in the first place! Even when they do the work correctly, it is far more helpful for the student herself, her peers, and teacher assessment if she is actually communicating her work and solution rather than merely recording calculations. This step-by-step communication can provide a sort of road map for student work and progress.

I gave the Taco Cart problem from Chapter 2 (page 25) to a class of eighth-grade Integrated Math students (http://threeacts.mrmeyer.com/tacocart/). Remember, this problem asks students which of two friends would reach the taco cart first, and in the communication and implementation phase, I asked students to let the friends know which route would be faster, and why. Let's consider one student's work:

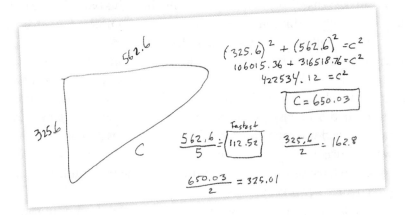

Even though this student started off correctly and understood the basic calculations, he lost track of his objective and stopped short of completing the problem. He clearly understood the basic mathematics in the problem, but did not communicate and implement the solution. He did not identify the meaning of his calculations, and did not make the connection back to the original problem. If he had made brief notes on each step, and been redirected back to the problem itself, he might have been able to complete the work and implement the solution properly, explaining to the friends which route would get them to the taco cart most quickly.

The following student's work on the Taco Cart problem uses a few more labels for each step, includes units in calculations, and keeps better track of her objective. At the end, she reaches an appropriate conclusion and communicates that solution in a clear, brief statement.

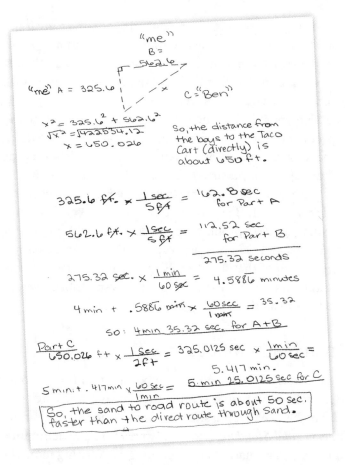

This student also expresses her solution using "friendly" numbers (using minutes and seconds rather than merely seconds, and stating that one route takes about 50 seconds less than another), reinforcing the fact that she understands the objective and can express the solution clearly. If this student were to look back on the problem, it would be clear what she was doing in each step, and she could more likely replicate her work on a similar problem. If her work were displayed for her fellow students, they would be able to follow her calculations, reasoning, and conclusions. In addition, the teacher could better assess her understanding, and if she had made an error, it would be easier to understand which concepts were causing her trouble.

➤ Communicating the Solution to Peers

Too often, both students and teachers see the final product of a modeling problem as the ultimate goal for getting a good grade, missing the point that the work itself can enhance everyone's understanding.

Students writing merely to get a good grade presume a great deal of prior knowledge on the part of their audience (the teacher) and may limit their explanation and justification. But students who must explain their solutions and justify their reasoning so that their classmates understand quickly become aware of the need to be clear and specific.

When students must justify their reasoning to their peers, particularly when those peers question that reasoning, everyone develops a deeper understanding of the mathematics involved. And when students see a problem approached from different perspectives, using different tools and methods, and hear a problem explained in different ways, everyone gains access to new ideas (Hatano and Inagaki 1991; Boaler and Humphreys 2005; Boaler and Staples 2008). When students see a problem approached from different perspectives, using different tools and methods, and explained in different ways, they gain a deeper understanding of that problem. Even when they have reached a correct solution, students can better understand a given problem when they see it explained, perhaps in a different way, by their peers.

➤ Writing in Mathematics

Writing in mathematics can become an important tool throughout problem solving and modeling, particularly when used to communicate and implement the solution. The NCTM emphasizes the importance of writing in mathematics in its *Principles and Standards*, stating, "Reflection and communication are intertwined processes in mathematics learning. . . . Writing in mathematics can also help students consolidate their thinking because it requires them to reflect on their work and clarify their thoughts about the ideas" (2000, 61). Writing about mathematics can enable students to analyze, compare and contrast, and synthesize information and data. It requires students to think about their own thinking and reasoning, focus on what is most important, and internalize and construct meaning out of their work (Kennedy 1980). The process of communicating the solution in writing is necessary not only to share ideas with others, but for the students to gain a deeper understanding of the mathematics themselves.

Writing to Help Teachers Assess Understanding

When talking with math teachers about using writing in their classrooms, I have heard many balk at the idea. I assure them (and any of you with similar concerns) that reading students' writing about math can be interesting and informative. I enjoy reading my students' thoughts and ideas about mathematics and the conclusions they draw from the solutions to modeling tasks. When students discuss and apply their solutions, they continue to engage the problem, make connections to other mathematics and real life, and gain a depth of understanding they did not have when they initially reached the solution. Writing in mathematics can also "help students consolidate their thinking because it requires them to reflect on their work and clarify their thoughts about the ideas developed in the lesson" (NCTM 2000, 61). In addition, we as teachers gain valuable insight into student reasoning and understanding that we cannot get from numerical solutions alone.

Joan Countryman (1992) identifies the assessment process as "an ongoing conversation between teacher and students about mathematical ideas," adding, "Writing in mathematics can be a significant part of this effort" (76). She identifies some teachers' questions about student understanding that can be answered through students' writing about the mathematics they are doing:

- Do students use math to make sense of complex situations?

- Can they formulate hypotheses?

- Can they organize information?

- Are they able to explain concepts?

- Can they use computation skills in context?

- Do they use mathematical language appropriately?

- Are they confident about using mathematical procedures?

Let's look at some student work through the lens of Countryman's questions. A group of eighth graders had been reading *Gulliver's Travels* in their English class. They came to class one day and found a giant's footprints on the sidewalk outside. Their task was to estimate the height of the giant, based on the size of his footprint.

At the conclusion of the class period, the students were asked to summarize their work and explain their reasoning. Here is one student's response:

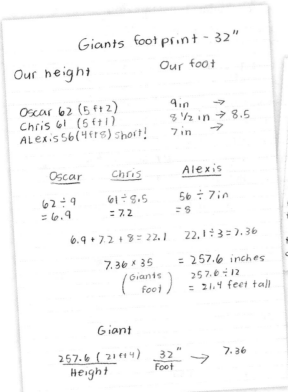

Giants footprint - 32"

Our height   Our foot

Oscar 62 (5 ft 2)   9 in   →
Chris 61 (5 ft 1)   8 ½ in → 8.5
Alexis 56 (4 ft 8) short! 7 in   →

Oscar   Chris   Alexis

62 ÷ 9   61 ÷ 8.5   56 ÷ 7 in
= 6.9   = 7.2   = 8

6.9 + 7.2 + 8 = 22.1   22.1 ÷ 3 = 7.36

7.36 × 35   = 257.6 inches
(Giants   257.6 ÷ 12
 Foot)   = 21.4 feet tall

Giant

257.6 (21 ft 4)   32"   →   7.36
Height   Foot

We predicted that the giant would be 224 inches tall because Oscar heard that people are 7 times taller than their foot size. The first thing we did was measure the giant's footprint. It was 32 inches. The next thing we did was measure all our feet and our height. We divided each person's height by the size of their foot. We added all of these together and divided by three. We got the number 7.36. We then multiplied this number (7.36) by the size of the giants foot (35 inches). By doing this we figured out the Giant was 257.6 inches tall. We divided this By 12 inches to get 21.4 feet tall. We think our prediction was close to the answer we got at the end.

- **Do students use math to make sense of complex situations?**
  This student used appropriate mathematics to quantify the relationship between height and foot length. Then, he used math appropriately to find average, scale factor, and apply the result to estimate the height of the giant.

- **Can they formulate hypotheses?**
  This student formulated a hypothesis. He used some prior knowledge ("Oscar heard that people are 7 times taller than their foot size") to predict the size of the giant.

- **Can they organize information?**
  Yes. This student organized data by group member, by foot size, and by height.

- **Are they able to explain concepts?**

  This student does a fairly good job of explaining concepts used. He was able to explain the steps taken, although he did not necessarily use appropriate mathematical language.

- **Can they use computation skills in context?**

  The student used almost all of the appropriate computation skills in context. (On the very last step, he appears to be checking his answer, but did not actually do so. Therefore, he did not notice that he had a transcription error, using "35 inches" for the giant's footprint, rather than 32 inches.)

- **Do they use mathematical language appropriately?**

  The student clearly understood most of the mathematics and used the appropriate mathematical language when describing operations; however, he did not include mathematical language regarding some important concepts: average or mean, ratio or proportion, or scale factor. When I asked the group to describe why they divided their heights by their foot lengths, one member told me, "Yeah, we didn't agree on what that number was called. Is it a ratio?"

- **Are they confident about using mathematical procedures?**

  This student appears to be confident in his reasoning and his solution. The only confusion expressed was, again, identifying exactly what he was doing when he divided height by foot length (and then multiplying the average by the giant's foot length). He was confident that it would work, but a little unclear about why.

This brief analysis of the written description helps clarify for me what the student understands and what he is still confused about. The information helps inform my further instruction. On this lesson, for instance, after each group had presented their work, we reviewed the different methods for solving the problem and the various solutions (and whether they made sense), and students discussed the strategies that made the most sense to them. We reviewed the mathematical language for the concepts students had used, and I encouraged them to use the proper language in the next task.

The addition of writing to communicate solutions can assist teachers' assessment of student understanding and inform future instruction. I could have relied solely on the numerical solution to the Giant's Footprint problem, but by asking students to write as well, I was able to find areas of confusion I might have otherwise missed.

Prompts for Writing Tasks

Many of our students will be uncomfortable and inexperienced at first when writing in the math classroom. They will have trouble reflecting on their work and reasoning. Often, good prompts can get students started in the writing task. Phyllis Whitin and David J. Whitin (2000) recommend pasting a set of prompts inside the cover of students' math journals. These prompts include:

1. What do you notice?

2. What do you find interesting?

3. What patterns do you see?

4. What surprises you?

5. What do you predict? Why?

6. What do your findings make you wonder?

7. What does this remind you of?

These prompts will serve as starting points for students who struggle with written reflection on mathematics. As students gain confidence in their writing, they will need fewer specific prompts, and will be able to write reflections on their work and explain their reasoning. As they progress through the writing process, and begin summarizing and communicating solutions, briefer prompts may be used to inspire student reflection. The following words, along with a brief question or two, might prompt students to better focus their written responses:

- Analyze

- Describe

- Evaluate

- Justify

- Reflect/Question

- Summarize

- Synthesize
  (Urquhart 2009, 16)

As you give students writing assignments in math, incorporate and define these terms so that your students become familiar with them. Let's look at an example of how we might walk students through these prompts on one problem:

Dakota has some dimes and nickels in her piggy bank. If she has exactly $.80, find all combinations of dimes an nickels Dakota could have.

Analyze: Break down the problem into parts. How would the number of nickels and the number of dimes contribute to the total value of the coins?

We will need to figure the amount of money using 5 cents per nickel and 10 cents per dime. We will multiply 5 times the number of nickels and 10 times the number of dimes.

Describe: Represent the problem in words. Write a description of how you would figure this out so that another student could understand and do the same thing.

We started with guess-and-check, multiplying the number of nickels by 5 and figuring out the number of dimes to make $.80. So, we started with 4 nickels (we changed it to 40 cents). That means we need 40 more cents, so 4 dimes.

So we figured that 20 + 60 = 80, so we needed 6 dimes. Then we kept trying other numbers until we found them all.

Evaluate: Determine the value of various combinations of dimes and nickels, explaining your work so someone else could understand.

We found a pattern and used it to find all the combinations. We made a table. We figured we could only have an even number of nickels:

NO. OF NICKELS	VALUE OF NICKELS	NO. OF DIMES	VALUE OF DIMES
0	0	8	80
1	5	—	—
2	10	7	70
4	20	6	60
6	30	5	50
8	40	4	40
10	50	3	30
12	60	2	20
14	70	1	10
16	80	0	0

Justify: Explain how you know your solution is correct.

We tried all the possible numbers. We knew that we couldn't have more than 8 dimes, and we couldn't have more than 16 nickels, and we tried all the numbers in between. When we figured out the nickels had to be even, we only had to try every other number. All of the combinations add up to 80 cents.

Reflect/Question: Think about a different way you could solve the same question, or think of a related question and see if you could solve it the same way.

We could have started with 0 dimes and 16 nickels, and then followed the pattern backward from there.

I don't think there would be as many combinations if we had used quarters and nickels or quarters and dimes. I wonder how many there would be.

Summarize: Restate your discoveries, and anything you might have learned from others' work, in your own words.

We discovered that it is easier to make sure you find all possible combinations if you go in order and find a pattern. It is also easier to start at zero of one coin and go up from there.

Synthesize: Combine strategies, ideas, and information from other sources. Put the ideas together in a general statement.

Some groups made an equation ($5x + 10y = 80$), some drew a picture, and some made a table, but they all mean the same thing. They all took the number of a coin times its value to get the total amount for the coin. Also, if you look for a pattern and follow it in order, you should be able to find all combinations. This should work on any kind of coins.

It is not necessary to use all prompts on every problem, but it may be helpful to use one of these prompts occasionally to get students comfortable with their meanings. In any case, practice with writing and writing prompts will help students to reflect on and express their knowledge, skills, and ideas, and share them with others.

When students have gained confidence and proficiency in writing about mathematics through prompts and directed questioning, they can begin to produce more elaborate written summaries and implement their solutions. We will examine some implementation strategies later in the chapter.

Modeling with Mathematics

Using Multiple Representations

As we introduce our students to mathematical modeling tasks, we should be encouraging them to represent the problem and communicate the solution to those tasks in multiple ways. Early in the process, students may not consider the various ways in which they might represent data or solutions, so it is important to remind them of the representations available to them. When presented with rich modeling tasks, students may become overwhelmed with information and decisions.

When students are communicating and implementing the solution to a modeling task, they begin to understand that different representations convey ideas in different ways and appeal to different learning styles. For instance, on the Old McDonald's Farm task in Chapter 4, we saw that different students approached the problem using different representations. Some were better able to understand the problem and work toward a solution using pictures, while others used tables or equations. In the same way, different students will better understand the communication of the solution through different representations. When we encourage students to communicate their solutions in a variety of ways, we help them increase their understanding of the problem they have solved, as well as increasing the opportunity for other students to better understand the problem from different viewpoints. As stated in NCTM's *Principles and Standards* (2000), "The ways in which mathematical ideas are represented is fundamental to how people can understand and use those ideas. When students gain access to mathematical representations and the ideas they represent, they have a set of tools that significantly expand their capacity to think mathematically" (67).

I presented the Thanksgiving Dinner Seating problem to an eighth-grade algebra class:

> Patty Pilgrim was planning Thanksgiving dinner for 24 guests. She rented square tables that seated one on each side. How many tables would she need to seat the guests?
>
> When Patty received the tables, she realized she did not want the guests seated at separate tables, so she began pushing them together like this:

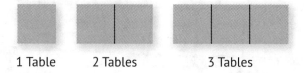

1 Table   2 Tables   3 Tables

When the tables are pushed together, what happens to the number of guests that can be seated? Describe any patterns you see and use your patterns to determine:

How many guests could be seated at 6 tables pushed together?

How many tables would Patty need to seat her 24 guests?

Write a generalization you could use to find the number of guests seated at *any number of tables.*

Students viewed the problem, set up the problem, and solved the problem in a variety of ways using a variety of representations. They then communicated their solutions using those multiple representations. I encouraged them to describe the patterns they found in words, tables, pictures, graphs, and equations.

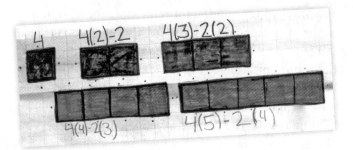

This group viewed, set up, and solved the problem by counting 4 guests per table, less 2 guests for each "crack" (where the tables were pushed together). They were able to demonstrate the way they reached their final expression, $4n - 2(n - 1)$, through the sketch and the numerical expressions on each group of tables.

Another group represented their reasoning and solution to this problem like this:

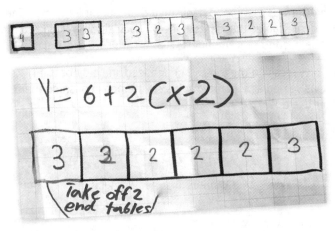

They also communicated their solution and equation using a sketch, but their reasoning was different from the first group's. Their reasoning is communicated clearly through their sketch and written explanation, showing that they counted 3 guests on each of the end tables (making the 6 in their equation), and then added 2 guests for every table in between ($x - 2$). The picture, combined with the written explanation, made this solution clear to the teacher and the other students and also helped clarify the connection between the equation and the concrete problem for the students themselves.

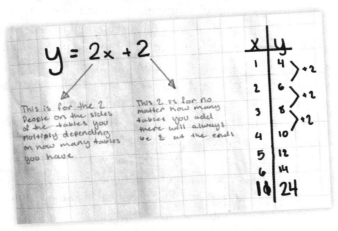

A third group communicated its solution and reasoning using a table of values and a written explanation. They demonstrated the constant rate of change on the table ($+2$), and then explained in writing that the slope of 2 represented the 2 people on the sides of each table, while the constant of 2 represented the 2 people who would always be on the ends of the tables. I asked this group to add a picture to their final product, reminding them that, although they understood the connection between their equation and the problem, the picture might make it clearer to their classmates. They added the following:

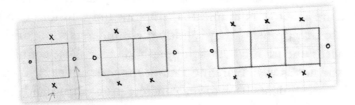

When students presented their results in a variety of ways, they found that they were not only providing a clear explanation of their reasoning and solution, but had a better understanding of the connection between the equation, the picture, and the actual problem they were solving (tables and guests).

It is apparent from the student work that students view this problem, reason about it, and solve it in a variety of ways. In the old days, my goal would have been to have all students come up with the same equation: $y = 2x + 2$, because that is clearly the "right" answer. As I have gained experience with this sort of problem, it is exciting to see that students can solve the same problem in a variety of ways, and that we can validate their reasoning without insisting on the one right answer. The range of expressions and equations reached by the students are perfectly, mathematically valid for this problem.

When I had students in this class present their solutions, we started with a peer review session in which one group presented to another group, who used a rubric to assess their peers' work. Upon initial observation of the other group's work, several students commented, "You got it wrong. Your equation is wrong." These students had noticed immediately that the other group's equation was different from their own and assumed it was incorrect. I encouraged the presenters to prove to the other group that their solution was different but correct, using their pictures, tables, and explanations. Students are accustomed to all getting the same, "right" answer in math class, so were confused when they saw several different equations that worked. Justifying and verbally explaining their work to their skeptical classmates added yet another layer to the level communication of the solution.

The Thanksgiving Dinner Seating problem has students start on the bottom rung of the ladder of abstraction: the concrete problem of tables and guests. As the students construct a model and solve the problem, they work at the representational level with manipulatives, pictures, tables, and graphs. Eventually, they represent their solutions abstractly by finding a generalization and an equation. The real depth of understanding occurs, however, when they work their way back down the ladder. By constructing a variety of representations to communicate their solutions, they make connections between the abstract equation, the picture and table, and the very real initial problem of tables and guests. When they return to the concrete problem and are able to communicate their work and solutions to others with respect to that problem, they solidify their understanding of the mathematical concept underlying their solutions and make those connections for their classmates as well. Bringing a modeling task full circle by posing a question brings students back to the concrete problem at hand. On the Thanksgiving Dinner Seating problem, for instance, students sometimes think they are finished when they come up with an explicit expression or equation that they can use to find the number of guests seated at any number of tables. They may need to be reminded to communicate to Patty

Modeling with Mathematics

Pilgrim the number of tables she will need for her 24 guests—the original problem posed.

Implementing the Solution

Whenever possible, students should physically implement the solution to rich modeling tasks in a real, concrete way. Although some problems cannot be demonstrated in a real way (like Old MacDonald's chicken and goat problem), students should put their work into action whenever possible. If some tasks have time or material constraints, it may be possible to replicate the problem on a smaller scale. At the least, students should implement the solution through a written recommendation, proposal, or explanation.

On the Taco Cart problem, for instance, you might ask the students to write a letter to the friends recommending the fastest route to the taco cart. You may show Dan Meyer's Act 3 video, with the overhead view of the boys "walking" to the taco cart. Your students might even try to replicate the scenario on the playground or field by walking slowly through the grass or sand, and more quickly on the blacktop for some given triangular path. For the Thanksgiving Dinner Seating problem, you might ask the students to write a note to Patty Pilgrim explaining the number of tables she will need for her party, or sketch a model explaining their table setup. For the Road Trip task (see page 6), students may implement their solutions by writing a letter to Mr. Wu, describing their recommended routes and the sights he might see along the way. On the Kicker Ramp problem (page 30), students may test the feasibility of their models using scale-model ramps and toy skateboards.

Whenever students can actually implement their solutions, they continue to learn and actively engage in the mathematics even after the problem is "solved." The grand finale of the Barbie Bungee problem (page 61), for instance, provides an exciting test of the students' work. When students have completed their calculations and provided a general plan or formula for finding the number of rubber bands needed to make a safe, exciting bungee jump, it is time to test their conclusions. For this stage in one classroom, I gave students a specific height from which we would have Barbie jump (in this case, the top of a ladder), and asked them to attach the appropriate number of rubber bands. We then took the Barbies and the ladder outside, and had them "jump" one by one. Some groups were rewarded by the successful jumps (Barbie approached the concrete below but bounced up in time), while other groups realized they had better adjust their models as Barbie's head hit the concrete!

Working with the 1,000 Paper Cranes Problem

On the 1,000 Paper Cranes problem, students are challenged to communicate the solution in creative ways. Their final step requires them to persuade the principal that their group's calculations are accurate and that they have determined the best plan for creating, producing, and displaying the cranes. In a recent class, the school principal attended the students' final presentation of results. This was the first modeling task the students had worked on in this class, and their experience communicating the solution was limited. Most groups created a visual display, such as a poster, to aid in their presentation to the principal and convince her that their calculations were correct.

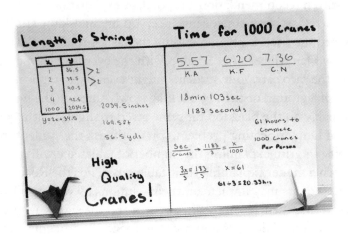

This group demonstrated to the principal that they had properly calculated the length of the string of cranes, as well as the time required to produce 1,000 cranes. As other groups were presenting, this group realized that their calculation of "61 hours to complete 1,000 cranes" did not make sense when compared with other groups' results. They quickly reassessed their model, and figured out that their calculations were for one person (rather than the entire group) to complete the work. You may notice that they then hastily wrote in "per person" on the poster and then added the line "61 ÷ 3 = 20.33 hours" at the bottom of the poster! Rather than being unhappy that the students had missed this calculation at the beginning, I was gratified to see that they were able to learn from the reasonableness of others' solutions, reconsider their work, and adjust their model and their solution before their presentation. This was a great example of the power communicating the solution to a problem. While some teachers might be hesitant about the time commitment involved in this stage of modeling, this example demonstrates the value of this step.

Not all of the solutions in this class were communicated on posters. Some students created a video "commercial," presenting their results and attempting to convince the principal of their reliability and work ethic. The video also showed the area of the school where the cranes would be best displayed. Another group made a multimedia presentation with software that displayed their calculations of time required to produce the 1,000 paper cranes, as well as the length of their string(s) of cranes. One student wrote a business proposal letter to the principal (and read it to her in class). Because this class was relatively inexperienced in this stage of the process, the presentations ranged in their ability to "communicate and implement the solution." Upon reflection at the conclusion of the presentations, students made comments such as:

"We were really confident about our calculations, and I think that made it easier for us to make our case to the principal."

"When other groups started presenting, we realized we must have made a mistake in our calculations. We figured it out and fixed it before we presented."

"I wish we had included a picture of the string of cranes on our poster. It would have made our work clearer when we presented it. I don't think the principal understood our equation!"

"Even though I know our results made sense, other groups' posters made it easier to understand what they did."

"I got some really good ideas from the other groups. Our next presentation will be better."

"The presentation part of the project brought all the other parts together. When we tried to explain it to the class and the principal, even I understood better how all the parts fit together."

This phase of the modeling process—communicating the solution—was a powerful one for these students. They understood the importance of presenting their reasoning in different ways and providing clear explanations of their work. Viewing the variety of work of other students further clarified their own work and reasoning, and they learned that they needed to make explicit connections between the different representations. The process of communicating their solutions reinforced their understanding and reasoning and challenged them to make their reasoning clear to others.

The implementation part of the 1,000 Paper Cranes problem was included in the communication of the solution. By presenting their proposals to the principal through writing, presentation software, or video, the students implemented their solutions and received feedback from the principal. Further implementation of this problem is also particularly effective. Even though the students will probably not have created 1,000 cranes, they can display their scale models in the classroom, the cafeteria, or the school office. This final implementation phase makes a very clear connection between the mathematics and the concrete problem, and also creates a beautiful visual display!

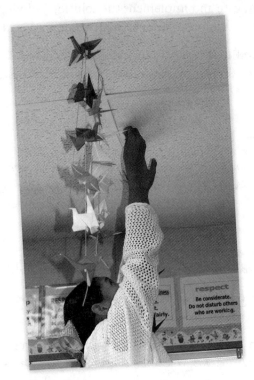

Conclusion

It is important to understand that calculating the solution is not the end of the modeling task. Too often, we teachers view the communication and implementation stage of the modeling process as superfluous or a waste of time. When students work beyond the solution, they develop a deeper understanding of the task, and when they must clearly convey their calculations and reasoning, they clarify their thinking. Further, when they employ multiple representations to communicate, they make deeper, abiding mathematical connections—for themselves and for others.

Assessing Mathematical Modeling

Most teachers waste their time by asking
questions that are intended to discover
what a pupil does not know, whereas the
true art of questioning is to discover
what the pupil does know or is capable of
knowing.

—ALBERT EINSTEIN

The recent history of education in the United States has featured a focus
on accountability and standards-based assessment. Because of various
reform efforts in the last thirty years (most notably in the No Child Left
Behind Act), the whole notion of assessment has been under a microscope.
Despite the bad taste some of these reform efforts have left in the mouths of
many educators, *assessment* is not a dirty word. Good assessment properly used
can set and place a focus on learning goals for teachers and students, inform
students and teachers about what students know and what they need to know,
give students feedback about performance and understanding, actively involve

students in the learning process, and inform teachers about the direction of further instruction.

Let's begin with an important clarification: *Assessment* is *not* synonymous with *testing*. Although testing is a part of assessment, the biggest, most informative, most important work of assessment happens in classrooms daily. Informal assessment involves the continuous task of listening, questioning, observing, and interacting with students (Bush, Leinwand, and Beck 2000). When we talk about assessment, we must consider both this informal type of assessment as well as the more formal assessment we do using quizzes, tests, and other tasks.

Goals of Assessment

The NCTM *Principles and Standards for School Mathematics* (2000) asserts that, "Assessment should support the learning of important mathematics and furnish useful information to both teachers and students" (22). Good assessment used properly can not only inform a teacher at the conclusion of instruction of what students know and are able to do, it can and should also be used as an integral part of daily practice in the classroom. Research demonstrates that when assessment is used in this way, student learning is enhanced and students demonstrate significant learning gains (Black and Wiliam 1998a, 1998b).

When we assess modeling in particular, we are focusing on the process as well as the product. Modeling problems, especially when students are initially asked to use them, can be somewhat daunting. Students often do not know where to begin, sometimes get stuck in the middle of the process, sometimes go off in an unnecessary direction, and often believe they are finished before they have addressed all the parts of the problem or properly demonstrated their understanding. Good assessment techniques and tools can address these modeling challenges and add to students' depth of understanding during the modeling process rather than merely pointing out their errors at its conclusion.

Informal Assessment

Informal assessment takes place every day in the classroom. One of my mentors, Dr. Judith Jacobs (2011), pounded this assessment acronym into my head: ABWA! Taken from a business management model, it means Assess By Walking Around. This informal assessment is particularly important when students are engaged in modeling tasks. I have highlighted several student-teacher interactions in previous chapters in which teachers are assessing student understanding, asking guiding questions, and adjusting instruction merely as a result of "walking around" the classroom.

In researching teachers' understanding of mathematical modeling, I found a fairly common misconception: Many teachers believe that modeling involves presenting students with a task and then leaving them to their own devices to tackle this task. This misperception frustrates teachers, who believe that their job involves helping their students learn! The truth is, when students are working on a rich modeling task, the teacher's role is vital. Yes, we want students to develop mathematical autonomy—to access methods, ideas, and tools to help them solve the problem themselves—but we must also engage in ongoing instruction to get them there. Part of this ongoing instruction involves formative assessment—evaluating student understanding, progress, misconceptions, and needs—and it occurs while we, the teachers, are walking around, listening and talking with students.

In a recent class period, students were working on the Barbie Bungee problem (page 61), and I noticed that the students in one group had all written, "Barbie falls 1 meter for every 12 rubber bands." While I understood this might have been a good beginning, I was concerned that they were applying proportional reasoning (incorrectly) to the problem. I asked them, "Okay, so how many rubber bands would you need for a 2.5-meter bungee jump?" I left them to struggle with this problem for a few minutes, and when I returned they proudly gave me the answer: "30 rubber bands!" They told me that they had multiplied the 12 rubber bands by 2.5 (displaying proportional reasoning, without writing a proportion). I knew they had not considered Barbie's height as a constant, and asked them to try out their answer. They tested the Barbie "jump" from a height of 2.5 meters with 30 rubber bands and quickly discovered (to their dismay) that their calculations could not be correct: Barbie was nowhere near the ground! I encouraged them to work on their calculations and, perhaps, draw a picture to see if they could find their error. After some discussion and consideration of their sketch, they were able to identify the error in their reasoning and correct their work (changing their answer to 38 rubber bands, because they could only count Barbie's height once, rather than 2.5 times). If I had merely assessed student understanding at the conclusion of their work, I would have missed the opportunity for students to correct their errors and deepen their understanding of the task. I might have learned that they incorrectly identified this as a proportional task, rather than a linear function, but they would not have learned anything in the process. By assessing and reassessing progress throughout the task, I could identify what they did know and direct and challenge them to test their conclusions and adjust their calculations. I gathered information about areas of student confusion, and they gathered information about when to properly apply (and not apply) proportional reasoning. This interaction led to a powerful discussion the next day about the characteristics of this task and why proportional reasoning was not appropriate. The informal ABWA contributed to student understanding for the entire class, and to a richer comparison of proportions and linear equations than I would have had otherwise.

On another modeling task, a sixth-grade class was divided into collaborative groups, given a box of pasta (different groups were given different pasta boxes of varying size), and presented with the following task:

> A pasta company is designing new boxes for its pasta. Find the volume of your pasta box, and design another pasta box with the same volume. The box must be a rectangular prism. Show your work and explain your reasoning.
>
> Now, design another pasta box (rectangular prism) that uses *less cardboard* but can hold at least as much pasta. Create your box. Open your original box of pasta and test your new box. What did you find? If necessary, adjust your model.
>
> Create a presentation for the pasta company explaining how your newly designed box will save them cardboard (and therefore, money) while still holding at least the same amount of pasta.

These students had had some experience with volume and surface area of prisms, and were given a brief demonstration of box making with construction paper. During the first day of the task, students measured their boxes, calculated volumes, and used various methods to design and sketch their new, same-volume boxes. Most groups started with a guess-and-check method, but some began to understand that deconstruction of the dimensions into their factors would make their work easier. The teacher was able to assess student understanding, and redirect when necessary, by walking around and discussing the work with the students.

Teacher: Tell me what you're doing.

Student 1: We found the volume of the box is about 100.

Teacher: 100 what?

Student 2: Cubic inches.

Teacher: Sounds good. How can you make another box with the same volume?

Student 1: We're trying to find 3 numbers that multiply to 100.

Teacher: What have you come up with?

Student 2 (laughing): So far, 100 × 1 × 1, but that's a pretty weird pasta box!

Teacher: Keep working!

The teacher learned, through a few quick questions, that these students understood the basic calculations for volume, and that they were on the right track, finding three factors with a product of 100 cubic inches.

Another group was hard at work, but was already making boxes rather than performing calculations. The teacher approached this group and asked:

Teacher: What are you trying to do?

Student 1: We're trying to make a box that will hold all our pasta.

Teacher: It looks like you're using a lot of paper here! What have you discovered?

Student 1: We're gonna measure them to find the volumes.

Teacher: What is the volume of your first pasta box?

Student 2: 8 times 12 times 18. That's 1,728. (This group had measured in centimeters.)

Teacher: 1,728 what?

Students: Umm . . .

Teacher: Remember to record your units. Then, let's see if we can find a quicker way to find a volume of 1,728. How are we finding volume?

Student 2: Multiplying the 3 numbers together—length, width, and height.

Teacher: Okay, so let's see if we can find a way to multiply 3 other numbers to get a product of 1,728. Keep working.

These students put aside their box making for the time being, and with this redirection, started calculating (they were allowed to use calculators). Although the teacher stepped away, she was still attending to this group's work.

Student 1: Okay. What if we make the tall side 15 instead of 18?

Student 2: No, 'cause 15 doesn't go into 1,728. Maybe 16? (dividing on the calculator) Yeah. That goes in 108 times. What goes into 108?

Student 3: That's 9 times 12. Can we still use the 12?

Student 1: Why not? That's a different box. Try it. 9 times 12 times 16 is . . . (calculating). Yes! It's 1,728! That works. Let's draw it.

By assessing the students' understanding and misconceptions, the teacher was able to redirect them and keep them from moving too far down the wrong path (and save a lot of paper!). She listened to what they already knew (how to calculate volume) and reinforced the idea that it would be a better use of their time to find three numbers with a product of 1,728. She stood by to assess their ability to apply this knowledge, and was gratified to see them quickly get on the right track.

A third group was using more advanced mathematical reasoning.

Teacher: Show me what you're doing.

Student 1: We're finding the dimensions of a box with the same volume: 1,296 centimeters.

Student 2: Cubic centimeters.

Teacher: How are you finding the dimensions?

Student (showing the teacher her work): Well, our box is 6 centimeters by 12 centimeters by 18 centimeters. That's 1,296. So we broke down each number and multiplied them different ways.

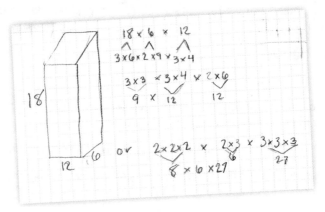

The teacher's assessment while the students were working gave her important information regarding their level of understanding. The students quickly realized, after some initial manipulation of the factors, that there were many possible dimensions that would produce the same volume. The teacher also saw that these students demonstrated fluency with factors and multiples, and with prime factorization.

So, merely by "walking around," we can receive important information about students' understanding, their areas of confusion, and common misconceptions. We can redirect the students before they travel too far in the wrong direction and advance their understanding by assessing what they already know. When we engage and direct students during the modeling process, we reduce student frustration and help students get "unstuck" by starting with what they already understand and guiding them as they proceed from there. This form of assessment is powerful because it increases student learning during the modeling process rather than merely informing the students about what they got wrong after they finished the process.

➤ Other Forms of Informal Assessment

Other methods of informal classroom assessment are particularly useful in longer modeling tasks. These include summaries, reflections, and exit slips. If students keep a math journal, ask them to summarize or reflect on the day's lesson. With practice, students learn to be specific and precise when recording responses to questions like "What is one thing you learned today?" and "What is one thing you are still confused about?" When they know in advance that they will be answering these questions at the conclusion of class, they attend to their questions (often writing them down) so they will remember them for their reflection. Many teachers use similar questions on an exit slip that students hand in on the way out of the classroom at the end of the period. Such summaries can be particularly helpful on multiday modeling tasks. It is helpful for students to recap their work for the day, pose any questions they may have, and think about what they might be working on the next day. Student reflections can also give the teacher a window into students' thinking.

At the end of the first day of the Pasta Box task, all the groups had calculated the volume of the original box as well as the dimensions of at least one other box with the same volume. Most groups also realized that the amount of cardboard needed required them to calculate surface area and had at least begun those calculations. In the last several minutes of the class, the teacher asked students to self-assess their understanding of the task by answering the following questions:

What did you accomplish today?

What did you learn today?

What do you need to do tomorrow?

What questions do you have?

In this class, several students recorded some variation of the following: "I'm not sure how I can build a smaller box that will hold all the pasta," so the teacher was able to introduce the next class by discussing this question.

Instead of performing calculations or giving specific directions on rectangular prisms, she encouraged students to experiment with the cardboard needed for different sizes of boxes, starting with the two (or more) boxes they found the day before. In addition, the teacher did a brief demonstration, asking students which of two cylinders (one tall and narrow, one short and squat) would hold the most water (they held approximately the same amount of water). This brief demonstration confirmed for students that it was possible to create very different containers with the same volume, and inspired them to start working again!

Reflections and summaries can be quick, informal assessments that provide important information for both the teachers and the students. Design the assessments so that students can reflect on their own understanding, and you can intervene during the learning process, rather than waiting until the end of instruction on a concept before discovering your students did not understand. Reflections, summaries, and exit slips can direct students to assess their own progress, understanding, and areas of confusion while being a quick and efficient way to review what students know and what they may be confused about. They allow teachers to adapt instruction for the group as a whole or give more directed guidance to particular students as they work on an extended task.

Formal Assessments

Although the majority of assessment work occurs informally during the course of a modeling task, formal assessment can also help students and teachers throughout the task. On a modeling task, the formal assessment can serve as both a formative and a summative assessment: that is, it can assist students in self- and peer assessment as they progress through a task, and can also serve as an instrument for effective and efficient final assessment at the conclusion of the task. Although the most frequently used summative assessment is a test, a test is not usually the best way to assess what students know and can do on a modeling task. Because of the more complex and often open-ended nature of modeling, most modeling tasks can be best assessed using a rubric.

➤ Effective Rubrics

Good teachers clearly identify and quantify what they want their students to know and be able to do *before they begin teaching the material.* They plan their instruction around learning goals. They also plan their assessments around these goals. A really good teacher also keeps her students in the loop: the students understand before, during, and at the conclusion of a lesson what it is they are expected to know and be able to do. I believe a good teacher does not keep these expectations secret from students! A very effective way to inform students *during the course of a modeling task* of what is expected from them is to make a rubric.

A rubric is a "coherent set of criteria for students' work that includes descriptions of levels of performance quality on the criteria" (Brookhart 2013). When students have a well-designed rubric as a reference while working on a modeling task, they are not in the dark about what they are expected to know and do. As a formative assessment tool, students can refer to the rubric for direction and reinforcement as they progress toward a final product. In addition, the rubric can be used for self- and peer assessment before it is used as a summative assessment.

When I started including performance tasks and projects in my classroom many years ago, I found that the students were engaged and excited by the change in routine and anxious to work together (in the days when "collaboration" was a rare occurrence in math classes). At the conclusion of the projects, however, I often found that the interesting, colorful posters showed little evidence of a depth of understanding of the mathematics involved. I had to reexamine what I was asking the students to do and how I was communicating my expectations to them. When rubrics came into vogue in math classes, I found an efficient, effective way to communicate my expectations and the project objectives to my students. At that point, I began to see a change in the projects in my classroom. When the students had clearly defined objectives, they were much more likely to concentrate on the mathematics rather than solely on the aesthetics of their presentations.

Rubrics have gained popularity in classrooms in the past couple of decades (Popham 1997). They are most properly used for more extensive performance tasks and are intended to focus on student understanding and the achievement of important skills and content. As Popham (1997, 73) asserts, "If performance tests are truly worth the effort that goes into creating and using them, we should evaluate them chiefly according to the contributions they make to students' skill mastery."

If you have been hesitant about rubric grading, the introduction of modeling tasks gives you an opportunity to incorporate this assessment tool into your classroom. Developing rubrics may be daunting at first but is well worth the effort! There are many good books and websites dedicated to creating a good rubric, and I invite you to explore them. When you begin using rubric grading on modeling tasks, you may want to start with a generic math performance rubric such as the newly revised Exemplars Standards Based Math Rubric from the Exemplars website (www.exemplars.com).

	PROBLEM SOLVING	**REASONING AND PROOF**	**COMMUNICATION**	**CONNECTIONS**	**REPRESENTATION**
Novice	No strategy is chosen, or a strategy is chosen that will not lead to a solution. Little or no evidence of engagement in the task present.	Arguments are made with no mathematical basis. No correct reasoning or justification for reasoning is present.	No awareness of audience or purpose is communicated. No formal mathematical terms or symbolic notations are evident.	No connections are made or connections are mathematically or contextually irrelevant.	No attempt is made to construct a mathematical representation.
Apprentice	A partially correct strategy is chosen, or a correct strategy for solving only part of the task is chosen. Evidence of drawing on some relevant previous knowledge is present, showing some relevant engagement in the task.	Arguments are made with some mathematical basis. Some correct reasoning or justification for reasoning is present.	Some awareness of audience or purpose is communicated. Some communication of an approach is evident through verbal/written accounts and explanations. An attempt is made to use formal math language. One formal math term or symbolic notation is evident.	A mathematical connection is attempted but is partially incorrect or lacks contextual relevance.	An attempt is made to construct a mathematical representation to record and communicate problem solving but is not accurate.
Practitioner	A correct strategy is chosen based on the mathematical situation in the task. Planning or monitoring of strategy is evident. Evidence of solidifying prior knowledge and applying it to the problem-solving situation is present. *Note: The Practitioner must achieve a correct answer.*	Arguments are constructed with adequate mathematical basis. A systematic approach and/or justification of correct reasoning is present.	A sense of audience or purpose is communicated. Communication of an approach is evident through a methodical, organized, coherent, sequenced, and labeled response. Formal math language is used to share and clarify ideas. At least two formal math terms or symbolic notations are evident, in any combination.	A mathematical connection is made. Proper contexts are identified that link both the mathematics and the situation in the task. Some examples may include one or more of the following: clarification of the mathematical or situational context of the task, exploration of mathematical phenomenon in the context of the broader topic in which the task is situated, noting patterns, structures, and regularities.	An appropriate and accurate mathematical representation is constructed and refined to solve problems or portray solutions.

	PROBLEM SOLVING	REASONING AND PROOF	COMMUNICATION	CONNECTIONS	REPRESENTATION
Expert	An efficient strategy is chosen and progress toward a solution is evaluated. Adjustments in strategy, if necessary, are made along the way, and/or alternative strategies are considered. Evidence of analyzing the situation in mathematical terms and extending prior knowledge is present. Note: The Expert must achieve a correct answer.	Deductive arguments are used to justify decisions and may result in formal proofs. Evidence is used to justify and support decisions made and conclusions reached.	A sense of audience and purpose is communicated. Communication at the practitioner level is achieved, and communication of argument is supported by mathematical properties. Formal math language and symbolic notation is used to consolidate math thinking and to communicate ideas. At least one of the math terms or symbolic notations is beyond grade level.	Mathematical connections are used to extend the solution to other mathematics or to a deeper understanding of the mathematics in the task. Some examples may include one or more of the following: testing and accepting or rejecting of a hypothesis or conjecture, explanation of phenomenon generalizing, and extending the solution to other cases.	An appropriate mathematical representation is constructed to analyze relationships, extend thinking, and clarify or interpret phenomenon.

This rubric, although not task specific, assesses performance tasks with a focus on NCTM's Process Standards as well as the Common Core Standards for Mathematical Practices.

There are many websites that provide free templates, sample rubrics, and teacher-created rubrics for your use. The Mathematics Assessment Resource Service (MARS) tasks at www.insidemathematics.org come with their own rubrics, and you can find templates and samples at sites such as Rubistar: http://rubistar.4teachers.org/index.php; teAchnology: www.teachnology.com/ ; and iRubric: www.rcampus.com/indexrubric.cfm to name a few. As you gain experience with generic rubrics and premade rubrics like the ones on the MARS tasks, you will gain confidence in your ability to design your own.

The key to making a good rubric is to clarify, before you begin the task, what you want your students to know and be able to do. Remember, for these larger modeling tasks, it is important to focus on the big ideas and goals, and since our best modeling rubrics are designed as formative as well as summative assessment tools, we want to write them in student-friendly language while still identifying specific criteria for each category. The format and specifics are up to you; remember, the goal is to clarify for your students what you expect from them and to make student assessment more meaningful and less time-consuming for you.

When I was creating the rubric for the Barbie Bungee task (page 61), I wanted students to demonstrate understanding and proficiency in the following areas:

Mathematical reasoning:

Demonstrate understanding of the linear nature of the problem.

Use data to develop a linear equation or explicit explanation of distance fallen relative to number of rubber bands used.

Mathematical content:

Demonstrate use of substitution in linear equations in order to find distance fallen with *any number of rubber bands* as well as number of rubber bands needed for a successful jump from *any given height*.

Multiple representations:

Demonstrate use of multiple representations (including tables, pictures, equations, graphs, and written explanations) to calculate and/or explain solutions to questions about Barbie Bungee

Written explanation:

Use a written explanation to explain the connection between the equation and the physical problem (Barbie, bungee, rubber bands). Explain to the bungee developer how to decide how many rubber bands are needed for any given jump.

Solution:

Demonstrate (by testing of an actual jump) that the solution is correct (as evidenced by safety and excitement of Barbie's jump).

Presentation:

Communicate results and recommendations in a clear, engaging, neat, and organized manner.

When I had clarified those goals and objectives in my own mind, I began creating a rubric that would demonstrate my expectations to my students. I used the goals and objectives as the categories, and decided on what would be required for each level of proficiency (expert, practitioner, apprentice, and novice).

CATEGORY	4: EXPERT	3: PRACTITIONER	2: APPRENTICE	1: NOVICE
Mathematical Reasoning	Demonstrates a clear and complete understanding of the linear nature of the problem. Constructs a correct equation or explicit explanation that can be used to find number of rubber bands needed for any given height. Variables are identified properly.	Demonstrates a good understanding of the linear nature of the problem. Constructs a correct equation or explicit explanation that can be used to find number of rubber bands needed for any given height. Variables are not identified properly.	Demonstrates an understanding of the relationship between height of jump and number of rubber bands. Some errors in calculation.	Demonstrates a poor understanding of the relationship between height of jump and number of rubber bands. Errors in calculation.
Mathematical Content	Demonstrates understanding and correct use of linear equation or explicit equation to find height of jump for any number of rubber bands as well as number of rubber bands needed for a jump from any height.	Demonstrates understanding and correct use of linear equation or explicit equation to find either height of jump for any number of rubber bands or number of rubber bands needed for a jump from any height.	Demonstrates limited understanding of use of linear equation or explicit equation to find either height of jump for any number of rubber bands or number of rubber bands needed for a jump from any height.	Demonstrates limited understanding of use of linear equation or explicit equation to find either height of jump for any number of rubber bands or number of rubber bands needed for a jump from any height.
Multiple Representations	The work is presented in multiple ways including: sketches, tables, graphs, written explanations, and equations. Connections between the representations are strong.	The work is presented in multiple ways including: sketches, tables, graphs, written explanations, and equations. Connections between the representations are adequate.	The work is presented in 2 or more ways including: sketches, tables, graphs, written explanations, and equations. Connections between the representations are weak.	The work is presented in 2 or fewer ways including: sketches, tables, graphs, written explanations, and equations. Connections between the representations are not made.
Written Explanation	Explanation is detailed and clear. It makes clear connections between the mathematics and the physical model.	Explanation is clear and makes connections between the mathematics and the physical model.	Explanation is somewhat difficult to understand, but includes critical components.	Explanation is difficult to understand and is missing several components OR was not included.
Solution	Barbie's bungee jump was extremely successful. She was very close to, but did not hit the ground!	Barbie's bungee jump was successful. She was close to, but did not hit the ground.	Barbie's bungee jump was nearly successful. She was either not close to the ground, or she touched the ground.	Barbie's bungee jump was unsuccessful. Barbie crashed into the ground!
Presentation	The work is presented in a neat, clear, and organized fashion that is easy to read and eye-catching.	The work is presented in a neat and organized fashion that is usually easy to read.	The work is presented in a fairly organized fashion but may be difficult to read or understand at times.	The work appears sloppy and unorganized. It is difficult to know what information goes together.

When I presented the students with the problem, I gave them time to consider the scenario, make some predictions, make a plan for data collection, and begin collecting data before I gave them the rubric. At that point, they were ready to use the rubric for reference and direction.

When my students were recently working on the Barbie Bungee project, I had this conversation with one group:

Student 1: I think we're done. Did we do everything?

Teacher: I don't know. Have you looked at the rubric? If you think you're finished, go ahead and use the rubric to assess your work. Let me know when you've come up with your final score.

I walked away for a few minutes, and when I returned, I found the students hard at work. They told me, "We aren't finished yet. We still need to write and explain our equation." When I referred them to the rubric, they had to reexamine their work and determine whether they had done all that was required of them, and I did not have to constantly remind them of the expectations.

At the conclusion of the task, students should be able to, within reason, assess their own final product using the criteria given on the rubric. In most cases, when I ask students to self-assess their work, they are harder on themselves than I would have been! Still, the act of comparing their work with the criteria on the rubric allows them to adjust and correct their work before final submission.

I have also found peer review using a rubric to be very effective. On more extensive modeling tasks, before the students present and submit their final products, I have them share their work with another group. The groups assess one another using the rubric, and the students have the opportunity to make corrections and additions before presenting to the entire class and/or submitting their work to me. This process is informative in a couple of different ways. First, the group being assessed receives feedback from their peers. This is often more palatable and understandable to students than feedback from their teachers. In addition, the group doing the assessing has another opportunity to review the expectations and reflect on their own work.

An example of peer assessment occurred recently in my classroom. One group was using the rubric to assess another group on the Thanksgiving Dinner Seating task (How many tables would be needed to seat 24 guests if you pushed the tables together?). One group initially wrote this explanation:

$$24 = 4(11) - 2(10)$$
$$44 - 20$$

When you multiply 4 times 11, 4 being the number of people, 11 being the number of tables. Then you subtract that by 2 times 11, 2 being the number of people, 10 being 1 less than the number of tables. You would get 44-20, which would equal 24.

The group doing the peer assessing asked some good questions of the presenters: How did you get the 11? What does the 4 mean in the equation? How would this help if we needed 30 guests or 100 guests? They let the presenting group know that this explanation was confusing to them, and they did not think it explained their reasoning. They marked the explanation as only Level 2 of 4 ("Apprentice") on the rubric, concluding that it met the criterion: "Explanation is somewhat difficult to understand, but includes critical components." Through the peer review session with the rubric, the presenting group understood that their explanation was insufficient and they were able to discuss it and adapt it before final submission. As an additional benefit, the group doing the assessing was required to match the other group's work to the rubric, discuss the criteria, come to a consensus, and explain their assessment to the presenting group. This process helped expand their own understanding and even reconsider their own explanation in terms of the rubric criteria.

Many modeling tasks are open-ended in that there are many ways to express the students' model (a linear equation, verbal description), and the data will be different for different groups (on the Barbie Bungee for instance, groups may have all small, all medium, or all large rubber bands, and some of the "Barbies" are actually other figures), so there are not necessarily "correct" answers. Still, with a rubric for guidance, the students understand what they are being required to do and how they will be assessed.

Even though modeling tasks often involve more in-depth investigations than typical math classwork, and they require some time for students to develop, revise, and test their models, the assessment of these tasks is relatively quick and easy with the rubric. A great deal of the summative assessment can be accomplished during the groups' presentations using the rubric as a clear guide. As each group begins its presentation to the class, I ask for their rubric. I make brief notes on the rubric and can usually mark the degree of proficiency on many of the categories as the group is presenting. On The Crow and the Pitcher task (page 26) (Wolf 2013a), the groups did their investigations (How many marbles would it take to raise the level of the water in a graduated cylinder to 100 milliliters?), and assessed themselves on their final product before presenting.

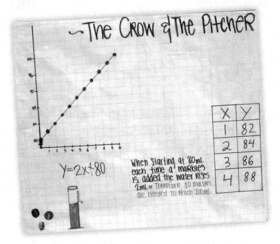

As this group presented to the class, I used the rubric with their self-assessment (the check marks) to assess them during their presentation (the boxes, notes, and final scores).

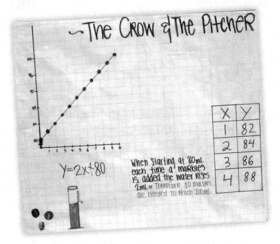

Reprinted with permission from *Illuminations: Resources for Teaching Math*, http://illuminations.nctm.org. Copyright © 2011 National Council of Teachers of Mathematics. Permission conveyed via Copyright Clearance Center, Inc.

Although the group assessed its own work as all "Expert," I showed them with very brief marks and notes where their poster was lacking and what they were missing.

I was able to complete the rubric assessment during each brief presentation (less than five minutes) by asking clarifying questions of each student to check for understanding. For instance, I was wavering between level 2 and 3 on the graph criterion because the axes were not labeled properly. I assessed it as a "minor problem" with the graph, because the graph was neat, organized, and correctly labeled with an appropriate scale, and students were able to quickly and easily tell me what each axis represented (and corrected the labeling on the graph at the conclusion of their presentation). So the rubric helped me clearly identify the expectations of the task to the students, guide their work as they conducted the investigation, help them to self-assess their work as they prepared their final product, make my final assessment quick and easy as they presented their results, and make the students' understanding of my assessment clear at the task's conclusion.

1,000 Paper Cranes Assessment

When I developed the 1,000 Paper Cranes task, I set forth the goals and objectives in advance. I knew I wanted my students to use multiple representations to solve the problems presented (How long will it take your group to make 1,000 cranes? How long will your string(s) of cranes be? Where/how would you display the string(s) of cranes at school?), to use appropriate tools and reasoning, and to be able to explain their strategies and reasoning. In addition, I wanted them to present their work in a way that would persuade the principal to "hire" their group for crane production. This was a project that was intended to reflect algebraic reasoning, particularly regarding proportional reasoning and linear functions, and I wanted students to demonstrate their knowledge of both. I created a rubric with these goals in mind.

For the ranking system (typically the top row in a table-style rubric), I chose to use a verbal description (advanced, proficient, basic, needs improvement), rather than a point ranking because I believed the terms represented a clearer indication of level of proficiency to students than mere numbers. Then I identified the criteria on which I wanted students to focus. In this case, I used: mathematical reasoning, mathematical concepts, strategy/procedures, multiple representations, explanation/discussion, and presentation. Finally, for each criterion, I identified specific evidence for each level of the ranking system. The resulting rubric could then be used as a guideline for students as they progressed through the task and as a scoring guide for me as I assessed the final product.

	ADVANCED	PROFICIENT	BASIC	NEEDS IMPROVEMENT
Mathematical Reasoning	Demonstrates complex math reasoning, including proportional reasoning and developing a linear equation.	Demonstrates effective math reasoning, including proportional reasoning and developing a linear equation.	Some evidence of math reasoning. Displays knowledge of proportions and reasoning about linear relationships.	Little evidence of mathematical reasoning.
Mathematical Concepts	Applies math concepts (including proportions and linear functions) appropriately and correctly.	Work shows substantial understanding of math concepts (including proportions and linear functions).	Work shows some understanding of math concepts (including proportions and linear functions).	Work shows limitation in understanding of math concepts (including proportions and linear functions).
Strategy/ Procedures	Demonstrates appropriate and efficient strategies to solve problems correctly.	Demonstrates use of effective strategies to solve problems correctly.	Demonstrates use of some strategies to solve problems correctly. May contain some errors.	Demonstrates limited use of strategies to solve problems. Errors are evident.
Multiple Representations	The work is presented in multiple ways including: sketches, tables, graphs, written explanations, and equations. Connections between the representations are strong.	The work is presented in multiple ways including: sketches, tables, graphs, written explanations, and equations. Connections between the representations are adequate.	The work is presented in 2 or more ways including: sketches, tables, graphs, written explanations, and equations. Connections between the representations are weak.	The work is presented in 2 or fewer ways including: sketches, tables, graphs, written explanations, and equations. Connections between the representations are not made.
Explanation/ Discussions	Explanation is detailed, makes connections between representations, and greatly adds to the understanding of the problems and their solutions. Discussion addresses all questions including time, length of string, and crane display.	Explanation is clear, makes some connections between representations, and adds to the understanding of the problems and their solutions. Discussion addresses all questions including time, length of string, and crane display.	Explanation is somewhat difficult to understand and makes few or weak connections between representations. Discussion addresses two or three questions including time, length of strong, and crane display.	Explanation is difficult to understand, and does not make connections between representations. Discussion addresses one or none of the question regarding time, length of string, and crane display.
Presentation	Presentation is neat, organized, engaging, and persuasive.	Presentation is neat and organized, but may be lacking in visual interest or persuasive effect.	Presentation is neat and organized, but lacks visual interest or is not persuasive in nature.	Presentation is lacking in neatness and organization, and it not persuasive in nature.

Modeling with Mathematics

At the very beginning of the investigation, I want students to be concerned with thinking about the problem, deciding on their initial approach, making predictions, and deciding what information and data need to be gathered, rather than focusing on specific rubric criteria. Once they have the opportunity to do some thinking and make some decisions and have begun gathering data, I give them the rubric to guide and assess their progress. While they are working, I am assessing their progress as well. When students believe they are finished, the groups present to one another in pairs. Each group assesses the other, using the rubric, and the groups discuss their conclusions with each other. Even at this stage, it is important to be "ABWAing" to be sure the peer groups are using the rubric as a reference, and giving valuable feedback. Finally, students are given time to address the concerns and complete the tasks before submission.

As students work through the final stages of the 1,000 Paper Cranes task (preparing the presentation), I remind them to consult the rubric. When they feel they are finished, I ask them to use the rubric to score themselves, marking on the rubric their level of proficiency on each of the criteria. As groups in one classroom were going through their individual assessments, I recorded the following comments:

> "Uh-oh. We only used two representations: a table and an equation. Maybe we should add a picture."

> "I don't think our poster is 'persuasive.' We need to add something to convince Ms. Raigosa to pick us."

> "Ms. Wolf, What does this mean—'Makes connections between representations'? Did we do that?"

> "Did we put the part about where to display the cranes? We need to put that in our explanation!"

By using the rubric as a reference, students were able to assess their own work and make corrections when needed. They asked for clarification if they were uncertain about the criteria, and they felt more confident about their final products when they submitted them.

Conclusion

It is important to remember that assessment is far more than scoring an exam or giving a grade. Assessment of modeling tasks takes place not only at the end but also during the course of the work. When we rely solely on summative assessment, we are getting a snapshot of what our students do not know at one moment in time: when we assess frequently while the work is going on, we can give students feedback about their performance and understanding and adapt our instruction to meet their needs (Fosnot 2014; Storeygard, Hamm, and Fosnot 2010). Good assessment of modeling tasks can increase depth of understanding as well as performance for our students, and can inform us, their teachers, about what they know and can do.

When you have decided in advance what you want your students to know and be able to do, and have clearly communicated the goals and expectations to your students, the task will progress more smoothly.

Doing Mathematics

8

You can teach a student a lesson for a
day; but if you can teach him to learn by
creating curiosity, he will continue the
learning process as long as he lives.

—Clay P. Bedford

I was taught, and therefore taught at first, in a very traditional manner. When I began my teaching career in the 1980s, I stood at the front of the classroom at the podium, I wrote on the blackboard or overhead projector, the desks were in rows, and the students worked independently. If the 1980s me walked into my classroom today, she would be very frightened! The students are seated in groups, I am typically walking around rather than standing at the front, and the classroom is far from quiet. But if the 1980s me stopped to observe and listen to the students, she would see them *doing* math—sharing ideas, strategizing, and talking about mathematics. I would tell her that it has been quite a transition for both the students and for me! It has not been easy, but it has certainly been worthwhile. Over the last several years, my classroom has been completely transformed—for the better! The inclusion of mathematical modeling tasks has changed the way I teach and the way my students learn for the better. I understand that some of these changes may seem overwhelming at first, but I would like to ease your concerns so that you become comfortable, and even excited about including modeling tasks in your classroom.

Mathematical modeling is a powerful practice that can engage our students and increase their understanding of mathematics. Modeling enables our students to *do* math, actively engaging in the work of mathematics. By doing math, our students are learning about more than the lesson at hand—they are preparing for a lifetime of learning and applying that mathematics in real life.

Teachers are often understandably hesitant to incorporate modeling tasks into their curriculum. In my research (Wolf 2013b), I have heard teachers express many reasons why they balk at including these tasks in their classrooms. These tasks can be time-consuming and are often open-ended, which charges teachers with exploring and understanding a multitude of approaches and solutions. Moreover, teachers may be required to assess students' written responses as well as their numerical ones, often including complex, in-depth presentations. Some teachers are reluctant to upset the apple cart—that is, to let go of some control in the classroom. Other teachers are uncomfortable with the potential noise and mess involved in some modeling tasks, and some are not convinced their students can proceed without a lot of specific instruction. Let me address some of these concerns.

"I don't have enough time."

Yes, a modeling task generally takes more class time than calculation problems. Remember the Problem Pyramid:

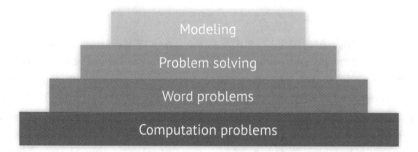

You are not going to use modeling tasks every day in your class. There is still plenty of room and time for traditional instruction and students' practice with concepts and computation. However, it is well worth the time investment to trade, perhaps, a day of "Do problems 2–34 even" for a modeling task now and then. You will still be able to address all of the standards, but you will produce mathematically autonomous thinkers in the process. Try one task each month or each grading period at first. I am confident that your experience with modeling and its impact on your students will encourage you to keep plugging along, using modeling more often and more confidently. Remember, not all modeling tasks have to be major, time-consuming endeavors. You can ease your class into modeling tasks with briefer, yet still rich tasks, and you will be rewarded by your students' engagement and depth of thinking.

"How can I possibly grade open-ended problems?"

The open-ended nature of many modeling tasks presents a challenge for many teachers. Some of the concerns expressed to me include, "How can I make an answer key for every possible student response?" or "How can I anticipate the methods or solutions students might use?" I can reassure you in a couple of ways. First, if you start with tried-and-true tasks with some direction (like the ones in this book, the MARS tasks, and those on the Illuminations website), you can use the teacher notes to plan in advance for many of the students' responses and questions. Also, once you begin using the modeling tasks, it will get easier every time. Step out into the unknown a little bit! I am always surprised (and gratified) to find new ways of looking at problems I have used many times before. Often it requires a back-and-forth discussion with the students, asking them to explain their work and reasoning (a good thing), and then testing their solution to be sure it works (also a good thing). In any case, open-ended tasks open doors to innovation and creativity and increase opportunities for the contributions of all learners. They also make for more exciting class presentations since students are not presenting or observing the same information over and over.

Many tasks will not require you to perform calculations in order to assess every possible solution; the students' presentations will often demonstrate their understanding of the mathematics and whether their answer is "correct." On the Barbie Bungee problem, for instance, the calculations were probably correct if Barbie gets close to the ground but doesn't hit her head on the final test jump! As students present the Pasta Box problem, they will be asked to show that they used less cardboard for their new box (by demonstrating understanding of surface area), and then demonstrate that their calculations actually work by pouring the pasta into the new box. On the 1,000 Paper Cranes problem, you will quickly get a feel for the approximate time it should take a group to produce the cranes, as well as the length of the string, depending on the configuration. Finally, if you are assessing work with a good rubric, you will see that it is possible to understand variations on open-ended problems without doing infinite calculations of every possible answer.

"I'm a math teacher. How am I supposed to grade writing?"

When we ask students to write in the math classroom, we are not grading the writing. We are using that writing to assess understanding and inform our future instruction. We are also using that writing to help students organize and articulate their reasoning and explain their thinking to others. We have discussed at length, and the research is clear, that writing is a powerful tool in the math classroom. Asking students to keep a math journal, explain their reasoning, and write reflections and summaries will increase the depth of knowledge of your students, but this can be a tool for teachers as well. You do not have to read every math journal entry every day, but it is very informative (and not very time-consuming) to flip through a particular day's summaries or responses when planning for the next day. Particularly when working with multiday modeling tasks, it is important to get the "big picture" about how students are progressing, and what we might have to review or reinforce before beginning the next day's task. Remember, when your students are working on a multiday task, you are doing very little "grading" throughout the task. This will give you the time to read through a few responses before moving on. At the end of the task, student reflections and summaries can also be informative regarding individual understanding and contribution to a group task.

Also, remember that you are reading brief explanations, questions, and reflections. You do not have to correct grammar and spelling if you are uncomfortable doing that, but at the very least, read for content. I can read through a class period's exit slips in less time than it takes to grade a page of computation problems. There are hurdles to overcome when asking students to write in the math class. If students are unaccustomed to this sort of reflection, it will seem very difficult at first. Keep trying. It will get easier! If students are English language learners, this task will be difficult as well, but I still encourage you to use your resources with these students and ask them to write. Depending on their level of English proficiency, you might allow them to respond in their native language or, if they have language assistance, dictate to an aide.

"I'm afraid of losing control of my class."

I attended a district math department meeting for secondary teachers several years ago when we were transitioning to the Common Core Standards for Mathematical Practices. We were asked to share how our classrooms would change with the incorporation of the Practices, particularly modeling, into our teaching. Nearly every teacher expressed the major changes in terms of classroom setup and management: "I have to train my students to move the desks quickly from rows to groups and back," or, "I don't know how I'll be able to stand the noise." I must admit I was a bit dismayed at these seemingly short-sighted concerns, but I realize they must be addressed before teachers can be asked to make major changes in the way they teach and the way their students learn. I also remember when I shared the same concerns.

If you crave order and quiet (like I used to), this may not be an easy transition for you, but I encourage you to try. The key to successful classroom management with modeling tasks is to set clear expectations and norms for cooperative groups. Some norms used in my classroom (developed with and by my students) include:

Respect each other's ideas.

Only one person speaks at a time—don't interrupt.

Everyone must participate and contribute.

Listen to one another.

Each person has a specific job.

Stay with your own group.

It is important to set norms early in the school year. If you do not yet have norms for group work, it might be advisable to first do one group project or activity. At its conclusion, the students might be more likely to recognize potential problems (students wandering away from the group, everyone trying to talk at once, one student doing most of the work) and address them in advance of the next group task, creating class norms in the process. As students are working on group tasks, it is often helpful to remind them of the norms they created.

I have come to appreciate, and even encourage, classroom "noise" (i.e., discussion), as long as it is productive noise related to the task at hand! Students will certainly need redirection and, sometimes, intervention, if they are not being productive. You will find your own norms, expectations, and comfort level with the classroom management issues involved in modeling.

Here are some further hints that may help you with classroom management when students are working in groups:

- *Change groups frequently:* Students can quickly fall into roles when working with the same classmates. Mixing it up a little allows students to contribute differently and demonstrate different strengths.

- *On a multiday task, set a clear goal about what should be accomplished by the end of each class period:* On a larger task, students may tend to meander a little, jumping from one thing to another. If you make it clear what the goal is for students to have accomplished by the end of the period, they are more likely to focus on the task at hand. Of course, things do not always go as planned, and some adjustments may need to be made. Still, having that goal in mind will help to keep students more focused and provide some direction.

- *Ensure that each group member has at least one specific "job," and that all members are contributing to the task. When presenting, each group member must present a part of the work, and it must be evident that each member understands the work:* The most challenging part of group work is trying to ensure that the work is shared among the group members. If each member has a mathematically valid job (not just, "You color the poster"), it is less likely that one student will do the majority of the work. Perhaps one member could create and present the graph, while another creates and presents the pictorial representation.

I try to promote an "All for one and one for all" mentality in the group modeling tasks. I remind students that every student in the group must completely understand the final product, how the group came to their conclusions, what they mean, and how they reached them. I remind them that, if one member does not understand, it is the responsibility of all other members to explain the work to him or her. When the group presents its results to me or to the class, they know I will ask questions of each person in the group about the process and the product. The entire group bears responsibility for any member who does not understand. (Obviously, there are exceptions here for the student who refuses to participate, is disruptive or uncooperative, or has been absent for part of the task. These students can be dealt with on a case-by-case basis.)

- *Wrap up each class period with a quiet, reflective time:* Keep an eye on the clock, and allow time for a brief recap of the day's task, and for students to reflect on the work of the day. This not only helps students look back and ahead to the next day, it brings the classroom back to a quiet, contemplative space before students move on. I have found that when the class period ends chaotically (putting things away, pushing desks around, turning things in), it is more difficult to get students on task for the next subject or the next class period.

"This sounds great for your students, but it will never work with mine."
This is the roadblock that most frustrates me. I want to shout from the rooftops, "Your students *can* do mathematical modeling tasks!" I have taught

and worked in classrooms from elementary school through graduate school. They can all do modeling tasks! I have taught the most struggling learners, and I have taught the most advanced learners. They are all capable of doing modeling tasks! I have taught students who have a long history of failure in mathematics, and who believe, "I can't do math." They *can* do powerful mathematics in modeling tasks! I have been continually rewarded by the successes and increased confidence of students who had convinced themselves they could not do math. I have heard time and time again, "I always thought I was stupid at math. Now I know I can do it." Give modeling a try. Have confidence in your students and in yourself. With rich tasks and your direction, they will achieve great things!

Works Cited

Ahmed, A. 1987. *Better Mathematics: A Curriculum Development Study Based on the Low Attainers in Mathematics Project.* London: HM Stationery Office.

Baruk, S. 1985. *L'âge du capitaine.* Paris: Seuil.

Bay-Williams, J. and S. Martinie. 2004. *Math and Literature: Grades 6–8.* Sausalito, CA: Math Solutions.

Black, P., and D. Wiliam. 1998a. "Assessment and Classroom Learning." *Assessment in Education* 5 (1): 7–74.

———. 1998b. *Inside the Black Box: Raising Standards Through Classroom Assessment.* London: Granada Learning.

Blumenfeld, P. C., E. Soloway, R. W. Marx, J. S. Krajcik, M. Guzdial, and A. Palincsar. 1991. "Motivating Project-Based Learning: Sustaining the Doing, Supporting the Learning." *Educational Psychologist* 26 (3–4): 369–98.

Boaler, J. 2001. "Mathematical Modelling and New Theories of Learning." *Teaching Mathematics and Its Applications* 20 (3): 121–27.

Boaler, J., and C. Humphreys. 2005. *Connecting Mathematical Ideas: Middle School Video Cases to Support Teaching and Learning.* Portsmouth, NH: Heinemann.

Boaler, J., and M. Staples. 2008. "Creating Mathematical Futures Through an Equitable Teaching Approach: The Case of Railside School." *The Teachers College Record* 110 (3): 608–45.

Brookhart, S. M. 2013. *How to Create and Use Rubrics for Formative Assessment and Grading.* Retrieved November 20, 2014 from http://www.ascd.org /publications/books/112001/chapters/What-Are-Rubrics-and-Why -Are-They-Important%C2%A2.aspx.

Buhl, D., M. Oursland, and K. Finco. 2003. "The Legend of Paul Bunyan: An Exploration in Measurement." *Mathematics Teaching in the Middle School* 8 (8): 441–48.

Bush, W. S., S. Leinwand, and P. Beck. 2000. *Mathematics Assessment: A Practical Handbook for Grades 6–8.* Reston, VA: National Council of Teachers of Mathematics.

Carpenter, T. P., E. Fennema, P. L. Peterson, C. P. Chiang, and M. Loef. 1989. "Using Knowledge of Children's Mathematics Thinking in Classroom Teaching: An Experimental Study." *American Educational Research Journal* 26 (4): 499–531.

Carpenter, T., M. Lindquist, W. Matthews, and E. Silver. 1983. "Results of the third NAEP Mathematics Assessment: Secondary School." *Mathematics Teacher* 76: 652–59.

Coerr, E. 1977. *Sadako and the Thousand Paper Cranes.* New York: Putnam.

Countryman, J. 1992. *Writing to Learn Mathematics: Strategies That Work.* Portsmouth, NH: Heinemann.

Deci, E. L., and R. M. Ryan. 1982. "Curiosity and Self-Directed Learning: The Role of Motivation in Education." In *Current Topics in Early Childhood Education,* ed. L. Katz. Vol. IV. Norwood, NJ: Ablex.

———. 1987. "The Support of Autonomy and the Control of Behavior." *Journal of Personality and Social Psychology* 53 (6): 1024.

Deci, E. L., R. J. Vallerand, L. G. Pelletier, and R. M. Ryan. 1991. "Motivation and Education: The Self-Determination Perspective." *Educational Psychologist* 26 (3/4): 325.

De Corte, E., et al. 1995. "Word Problems: Game or Reality? Studies of Children's Beliefs About the Role of Real-World Knowledge in Mathematical Modeling." Paper presented at the annual meeting of American Educational Research Association, San Francisco, CA.

De Jong, T., and W. R. Van Joolingen. 1998. "Scientific Discovery Learning with Computer Simulations of Conceptual Domains." *Review of Educational Research* 68 (2): 179–201.

English, L. D., J. L. Fox, and J. J. Watters. 2005. "Problem Posing and Solving with Mathematical Modeling." *Teaching Children Mathematics* 12 (3): 156.

Federal Education Budget Project. 2012. "No Child Left Behind: Overview." In *Federal Education Budget Projects,* ed. N. A. Foundation. Washington DC: New America Foundation.

Fosnot, C. 2014. *Determining What Children Know: Dynamic vs. Static Assessment.* Paper presented at the California Math Council, South Palm Springs, CA.

Greer, B. 1993. "The Mathematical Modeling Perspective on Wor(l)d Problems." *Journal of Mathematical Behavior* 12 (3): 239–50.

———. 1997. "Modelling Reality in Mathematics Classrooms: The Case of Word Problems." *Learning and Instruction* 7 (4): 293–307. doi: 10.1016/s0959 -4752(97)00006-6.

Hatano, G., and K. Inagaki. 1991. "Sharing Cognition Through Collective Comprehension Activity." From website psycnet.apa.org.

Hayakawa, S. I. 1940. *Language in Action: A Guide to Accurate Thinking, Reading, and Writing.* New York: Harcourt, Brace and Company.

Hiebert, J., and T. P. Carpenter. 1992. "Learning and Teaching with Understanding." From website psycnet.apa.org.

Hiebert, J., T. P. Carpenter, E. Fennema, K. Fuson, P. Human, H. Murray, D. Wearne et al. 1996. "Problem Solving as a Basis for Reform in Curriculum and Instruction: The Case of Mathematics." *Educational Researcher* 25 (4): 12–21.

Inoue, N. 2009. "The Issue of Reality in Word Problem Solving: Learning from Students' Justifications of 'Unrealistic' Solutions to Real Wor(l)d Problems." In *Words and Worlds: Modelling Verbal Descriptions of Situations,* ed. Verschaffel et al., 195–209. Rotterdam, Germany: Sense Publishers.

Jacobs, J. E. 2011. *A Winning Formula for Mathematics Instruction: Converting Research into Results.* Alexandria, VA: Educational Research Services.

Kaput, J. J. 1989. "Linking Representations in the Symbol Systems of Algebra." *Research Issues in the Learning and Teaching of Algebra* 4: 167–94.

Kaput, J., and R. Schorr. 2008. "Changing Representational Infrastructures Changes Most Everything: The Case of SimCalc, Algebra, and Calculus." In *Research on Technology and the Teaching and Learning of Mathematics: Cases and Perspectives Volume 2*, eds. G. Blume, and M. Heid, 211–242. Reston, VA: NCTM.

Kennedy, M. L. 1980. "Reading and Writing: Interrelated Skills of Literacy on the College Level." *Reading World* 20 (2): 131–41.

Krutetskii, V. 1976. *The Psychology of Mathematical Abilities in Schoolchildren*. Chicago: The University of Chicago Press.

Leinwand, S. 2009. *Accessible Mathematics: 10 Instructional Shifts That Raise Student Achievement*. Portsmouth, NH: Heinemann.

Lesh, R. 1981. "Applied Mathematical Problem Solving." *Educational Studies in Mathematics* 12 (2): 235–64.

Lesh, R., H. M. Doerr, G. Carmona, and M. Hjalmarson. 2003. "Beyond Constructivism." *Mathematical Thinking and Learning* 5 (2–3): 211–33.

Lesh, R., and R. Lehrer. 2003. "Models and Modeling Perspectives on the Development of Students and Teachers." *Mathematical Thinking and Learning* 5 (2–3): 109–29.

Lesh, R., J. Zawojewski, and G. Carmona. 2003. "What Mathematical Abilities Are Needed Beyond School in a Technology-Based Age of Information." In *Beyond Constructivism: Models and Modeling Perspectives on Mathematics Problem Solving, Learning, and Teaching*, 205–222, eds. Lesh, R., and H. Doerr. Mahwah, NJ: Lawrence Erlbaum Associates.

Lockhart, P. 2009. *A Mathematician's Lament: How School Cheats Us out of Our Most Fascinating and Imaginative Art Forms*. New York: Bellevue Literary.

Mason, J. 1989. "Mathematical Abstraction as the Result of a Delicate Shift of Attention." *For the Learning of Mathematics* 9 (2): 2–8.

Maslow, Abraham. 1966. *The Psychology of Science: A Reconnaissance*. New York: Harper & Row.

Mathematics Assessment Project. 2011. "100 People." *The Mathematics Assessment Project*. Retrieved March 12, 2014 from http://map.mathshell.org /materials/download.php?fileid=1046.

Meyer, D. 2011. "The Three Acts of a Mathematical Story." http://blog.mrmeyer .com/2011/the-three-acts-of-a-mathematical-story/.

Miller, L. D. 1991. "Writing to Learn Mathematics." *The Mathematics Teacher* 84 (7): 516–21.

National Council of Teachers of Mathematics. 2000. *Principles and Standards for School Mathematics*. Reston, VA: NCTM.

———. 2012. *Real World Math: Articles, Lesson Plans, and Activities for the Middle Grades*. Retrieved May 21, 2014, from www.nctm.org/publications/worlds/default.aspx.

National Governors Association Center for Best Practices. 2010. Common Core State Standards for Mathematics. www.corestandards.org/the-standards.

Noyce Foundation. 2012. "Performance Assessment Test: Mixing Paints— Grade 7." www.insidemathematics.org/assets/common-core-math -tasks/mixing paints.pdf.

Ormell, C. 1991. "How Ordinary Meaning Underpins the Meaning of Mathematics." *For the Learning of Mathematics* 11 (3): 25–30.

Pape, S. J., and M. A. Tchoshanov. 2001. "The Role of Representation(s) in Developing Mathematical Understanding." *Theory into Practice* 40 (2): 118–27.

Piggott, J. 2011. "Rich Tasks and Contexts." University of Cambridge. Retrieved February 24, 2014, from http://nrich.maths.org/5662.

Polya, G. 2008. *How to Solve It: A New Aspect of Mathematical Method.* Princeton, NJ: Princeton University Press.

Popham, W. J. 1997. "What's Wrong—and What's Right—with Rubrics." *Educational Leadership* 55: 72–75.

Presmeg, N. C. 1999. *On Visualization and Generalization in Mathematics.* Paper presented at the Proceedings of the Annual Meeting of the North American Chapter of the International Group for the Psychology of Mathematics Education, Cuernavaca, Morelos, Mexico.

Reeve, J., E. Bolt, and Y. Cai. 1999. "Autonomy-Supportive Teachers: How They Teach and Motivate Students." *Journal of Educational Psychology* 91 (3): 537–48. doi: 10.1037/0022-0663.91.3.537.

Rigelman, N. 2007. "Fostering Mathematical Thinking and Problem Solving: The Teacher's Role." *Teaching Children Mathematics* (February): 308–14.

Schiefele, U., and M. Csikszentmihalyi. 1995. "Motivation and Ability as Factors in Mathematics Experience and Achievement." *Journal for Research in Mathematics Education* 26 (2): 163–81.

Secondary National Strategy: Mathematics at Key Stage 4: Developing A Scheme of Work. 2007. *Secondary National Strategy.* http://webarchive.nationalarchives. gov.uk/20110202093118/http:/nationalstrategies.standards.dcsf.gov.uk/node /47454?uc=force_uj.

Seeley, C. L. 2014. *Smarter Than We Think: More Messages About Math, Teaching, and Learning in the 21st Century.* Sausalito, CA: Math Solutions.

Stefanou, C. R., K. C. Perencevich, M. DiCintio, and J. C, Turner. 2004. "Supporting Autonomy in the Classroom: Ways Teachers Encourage Student Decision Making and Ownership." *Educational Psychologist* 39 (2): 97–110.

Storeygard, J., J. Hamm, and C. T. Fosnot. 2010. "Determining What Children Know: Dynamic Versus Static Assessment." In *Reweaving the Tapestry: Models of Intervention in Mathematics,* ed. C.T. Fosnet, 45–69. Reston, VA: National Council of Teachers of Mathematics.

Swetz, F. 1991. "Implementing the Standards: Incorporating Mathematical Modeling into the Curriculum." *Mathematics Teacher* 84 (5): 358–65.

Swetz, F., J. S. Hartzler, and National Council of Teachers of Mathematics. 1991. *Mathematical Modeling in the Secondary School Curriculum*. Reston, VA: National Council of Teachers of Mathematics.

Turner, J. C., K. B. Warzon, and A. Christensen. 2011. "Motivating Mathematics Learning." *American Educational Research Journal* 48 (3): 718–62. doi: 10.3102/0002831210385103.

Urquhart, V. 2009. "Using Writing in Mathematics to Deepen Student Learning." Denver, CO: *Mid-continent Research for Education and Learning (McREL)*.

Verschaffel, L., and E. De Corte. 1997. "Teaching Realistic Mathematical Modeling in the Elementary School: A Teaching Experiment with Fifth Graders." *Journal for Research in Mathematics Education* 28 (5): 577–601.

Warwick, J. 2007. "Some Reflections on the Teaching of Mathematical Modeling." *Mathematics Educator* 17 (1): 32–41.

Webb, N. L. 2002. "Depth-of-Knowledge Levels for Four Content Areas. *Language Arts*. http://ossucurr.pbworks.com/w/file/fetch/49691156/Norm%20web%20dok%20by%20subject%20area.pdf.

Weber, K., I. Radu, M. Mueller, A. Powell, C. Maher. 2010. "Expanding Participation in Problem Solving in a Diverse Middle School Mathematics Classroom." *Mathematics Education Research Journal* 22 (1): 91–118. doi: 10.1007/bf03217560.

Whitin, P., and D. J. Whitin. 2000. *Math Is Language Too: Talking and Writing in the Mathematics Classroom*. Urbana, IL: National Council of Teachers of English.

Wolf, N. B. 2013a. "The Crow and the Pitcher: Investigating Linear Functions Using a Literature-Based Model." Available from NCTM's Illumintations: Resources for Teaching Math website. http://illuminations.nctm.org/Lesson.aspx?id=3667.

———. 2013b. *Teachers' Understanding of and Concerns About Mathematical Modeling in the Common Core Standards*. Ph.D. Doctoral dissertation, Claremont Graduate University, ProQuest Dissertations and Theses database (UMI No. 3590815).

Woodward, J., S. Beckmann, M. Driscoll, M. Franke, P. Herzig, A. Jitendra et al. 2012. *Improving Mathematical Problem Solving in Grades 4 Through 8. IES Practice Guide. NCEE 2012-4055*. Princeton, NJ: What Works Clearinghouse.

Yackel, E., and P. Cobb. 1996. "Sociomathematical Norms, Argumentation, and Autonomy in Mathematics." *Journal for Research in Mathematics Education* 27 (4): 458.

DATE DUE

PRINTED IN U.S.A.